D1024871

AI SUPERPOWERS

AI
SUPERPOWERS

★

CHINA,
SILICON VALLEY,
AND THE
NEW WORLD ORDER

Kai-Fu Lee

Houghton Mifflin Harcourt
Boston   New York
2018

For information about permission to reproduce selections
from this book, write to trade.permissions@hmhco.com or to
Permissions, Houghton Mifflin Harcourt Publishing Company,
3 Park Avenue, 19th Floor, New York, New York 10016.

hmhco.com

Library of Congress Cataloging-in-Publication Data
Names: Lee, Kai-Fu, author.
Title: AI superpowers : China, Silicon Valley, and the new world order /Kai-Fu Lee.
Description: Boston : Houghton Mifflin Harcourt, [2018] |
Includes bibliographical references and index.
Identifiers: LCCN 2018017250 (print) | LCCN 2018019409 (ebook) |
ISBN 9781328545862 (ebook) | ISBN 9781328546395 (hardcover)
ISBN 9781328606099 (international edition)
Subjects: LCSH: Artificial intelligence — Economic aspects — China. |
Artificial intelligence — Economic aspects — United States.
Classification: LCC HC79.I55 (ebook) | LCC HC79.I55 L435 2018 (print) |
DDC 338.4/700630951 — DC23
LC record available at https://lccn.loc.gov/2018017250

Book design by Chrissy Kurpeski

Printed in the United States of America
DOC 10 9 8 7 6 5
4500755475

To Raj Reddy, my mentor in AI and in life

CONTENTS

INTRODUCTION

One of the obligations that comes with my work as a venture-capital (VC) investor is that I often give speeches about artificial intelligence (AI) to members of the global business and political elite. One of the joys of my work is that I sometimes get to talk about that very same topic with kindergarteners. Surprisingly, these two distinctly different audiences often ask me the same kinds of questions. During a recent visit to a Beijing kindergarten, a gaggle of five-year-olds grilled me about our AI future.

"Are we going to have robot teachers?"

"What if one robot car bumps into another robot car and then we get hurt?"

"Will people marry robots and have babies with them?"

"Are computers going to become so smart that they can boss us around?"

"If robots do everything, then what are we going to do?"

These kindergarteners' questions echoed queries posed by some of the world's most powerful people, and the interaction was revealing in several ways. First, it spoke to how AI has leapt to the forefront of our minds. Just a few years ago, artificial intelligence was a field that lived primarily in academic research labs and science-fiction films. The average person may have had some sense that AI was about building robots that could think like people, but there was almost no connection between that prospect and our everyday lives.

Today all of that has changed. Articles on the latest AI innovations blanket the pages of our newspapers. Business conferences on leveraging AI to boost profits are happening nearly every day. And governments around the world are releasing their own national plans for harnessing the technology. AI is suddenly at the center of public discourse, and for good reason.

Major theoretical breakthroughs in AI have finally yielded practical applications that are poised to change our lives. AI already powers many of our favorite apps and websites, and in the coming years AI will be driving our cars, managing our portfolios, manufacturing much of what we buy, and potentially putting us out of our jobs. These uses are full of both promise and potential peril, and we must prepare ourselves for both.

My dialogue with the kindergartners was also revealing because of where it took place. Not long ago, China lagged years, if not decades, behind the United States in artificial intelligence. But over the past three years China has caught AI fever, experiencing a surge of excitement about the field that dwarfs even what we see in the rest of the world. Enthusiasm about AI has spilled over from the technology and business communities into government policymaking, and it has trickled all the way down to kindergarten classrooms in Beijing.

This broad-based support for the field has both reflected and fed into China's growing strength in the field. Chinese AI companies and researchers have already made up enormous ground on their American counterparts, experimenting with innovative algorithms and business models that promise to revolutionize China's economy. Together, these businesses and scholars have turned China into a bona fide AI superpower, the only true national counterweight to the United States in this emerging technology. How these two countries choose to compete and cooperate in AI will have dramatic implications for global economics and governance.

Finally, during my back-and-forth with those young students, I stumbled on a deeper truth: when it comes to understanding our AI future, we're all like those kindergartners. We're all full of questions without answers, trying to peer into the future with a mixture of childlike wonder and grown-up worries. We want to know what AI automation will mean for our jobs and for our sense of purpose. We want to know which people and countries will benefit from this tremendous technology. We wonder whether AI can vault us to lives of material abundance, and whether there is space for humanity in a world run by intelligent machines.

No one has a crystal ball that can reveal the answers to these questions for us. But that core uncertainty makes it all the more important that we ask these questions and, to the best of our abilities, explore the answers. This book is my attempt to do that. I'm no oracle who can perfectly predict our AI future, but in exploring these questions I can bring my experience as an AI researcher, technology executive, and now venture-capital investor in both China and the United States. My hope is that this book sheds some light on how we got here, and also inspires new conversations about where we go from here.

Part of why predicting the ending to our AI story is so difficult is because this isn't just a story about machines. It's also a story about human beings, people with free will that allows them to make their own choices and to shape their own destinies. Our AI future will be created by us, and it will reflect the choices we make and the actions we take. In that process, I hope we will look deep within ourselves and to each other for the values and wisdom that can guide us.

In that spirit, let us begin this exploration.

AI SUPERPOWERS

1

CHINA'S SPUTNIK MOMENT

The Chinese teenager with the square-rimmed glasses seemed an unlikely hero to make humanity's last stand. Dressed in a black suit, white shirt, and black tie, Ke Jie slumped in his seat, rubbing his temples and puzzling over the problem in front of him. Normally filled with a confidence that bordered on cockiness, the nineteen-year-old squirmed in his leather chair. Change the venue and he could be just another prep-school kid agonizing over an insurmountable geometry proof.

But on this May afternoon in 2017, he was locked in an all-out struggle against one of the world's most intelligent machines, AlphaGo, a powerhouse of artificial intelligence backed by arguably the world's top technology company: Google. The battlefield was a nineteen-by-nineteen lined board populated by little black and white stones — the raw materials of the deceptively complex game of Go. During game play, two players alternate placing stones on the board, attempting to encircle the opponent's stones. No human on Earth could do this better than Ke Jie, but today he was pitted against a Go player on a level that no one had ever seen before.

Believed to have been invented more than 2,500 years ago, Go's history extends further into the past than any board game still played today. In ancient China, Go represented one of the four art forms any Chinese scholar was expected to master. The game was believed to imbue its players with a Zen-like intellectual refinement and wisdom. Where games like Western chess were crudely tactical, the game of Go is based on patient positioning and slow encirclement, which made it into an art form, a state of mind.

The depth of Go's history is matched by the complexity of the game itself. The basic rules of gameplay can be laid out in just nine sentences, but the number of possible positions on a Go board exceeds the number of atoms in the known universe. The complexity of the decision tree had turned defeating the world champion of Go into a kind of Mount Everest for the artificial intelligence community — a problem whose sheer size had rebuffed every attempt to conquer it. The poetically inclined said it couldn't be done because machines lacked the human element, an almost mystical feel for the game. The engineers simply thought the board offered too many possibilities for a computer to evaluate.

But on this day AlphaGo wasn't just beating Ke Jie — it was systematically dismantling him. Over the course of three marathon matches of more than three hours each, Ke had thrown everything he had at the computer program. He tested it with different approaches: conservative, aggressive, defensive, and unpredictable. Nothing seemed to work. AlphaGo gave Ke no openings. Instead, it slowly tightened its vise around him.

THE VIEW FROM BEIJING

What you saw in this match depended on where you watched it from. To some observers in the United States, AlphaGo's victories signaled not just the triumph of machine over man but also of Western technology companies over the rest of the world. The previous two decades had seen Silicon Valley companies conquer world technology markets. Companies like Facebook and Google had become the go-to internet platforms for socializing and searching. In the process, they had steamrolled local startups in countries from France to Indonesia. These internet juggernauts had given the United States a dominance of the digital world that matched its military and economic power in the real world. With AlphaGo — a product of the British AI startup DeepMind, which had been acquired by Google in 2014 — the West appeared poised to continue that dominance into the age of artificial intelligence.

But looking out my office window during the Ke Jie match, I saw something far different. The headquarters of my venture-capital fund is located in Beijing's Zhongguancun (pronounced "jong-gwan-soon") neighborhood, an area often referred to as "the Silicon Valley of China." Today, Zhongguancun is the beating heart of China's AI movement. To people here, AlphaGo's victories were both a challenge and an inspiration. They turned into China's "Sputnik Moment" for artificial intelligence.

When the Soviet Union launched the first human-made satellite into orbit in October 1957, it had an instant and profound effect on the American psyche and government policy. The event sparked widespread U.S. public anxiety about perceived Soviet technological superiority, with Americans following the satellite across the night sky and tuning in to Sputnik's radio transmissions. It triggered the creation of the National Aeronautics and Space Administration (NASA), fueled major government subsidies for math and science education, and effectively launched the space race. That nationwide American mobilization bore fruit twelve years later when Neil Armstrong became the first person ever to set foot on the moon.

AlphaGo scored its first high-profile victory in March 2016 during a five-game series against the legendary Korean player Lee Sedol, winning four to one. While barely noticed by most Americans, the five games drew more than 280 million Chinese viewers. Overnight, China plunged into an artificial intelligence fever. The buzz didn't quite rival America's reaction to Sputnik, but it lit a fire under the Chinese technology community that has been burning ever since.

When Chinese investors, entrepreneurs, and government officials all focus in on one industry, they can truly shake the world. Indeed, China is ramping up AI investment, research, and entrepreneurship on a historic scale. Money for AI startups is pouring in from venture capitalists, tech juggernauts, and the Chinese government. Chinese students have caught AI fever as well, enrolling in advanced degree programs and streaming lectures from international researchers on their smartphones. Startup founders are furiously pivoting, reengineering, or simply rebranding their companies to catch the AI wave.

And less than two months after Ke Jie resigned his last game to

AlphaGo, the Chinese central government issued an ambitious plan to build artificial intelligence capabilities. It called for greater funding, policy support, and national coordination for AI development. It set clear benchmarks for progress by 2020 and 2025, and it projected that by 2030 China would become the center of global innovation in artificial intelligence, leading in theory, technology, and application. By 2017, Chinese venture-capital investors had already responded to that call, pouring record sums into artificial intelligence startups and making up 48 percent of all AI venture funding globally, surpassing the United States for the first time.

A GAME AND A GAME CHANGER

Underlying that surge in Chinese government support is a new paradigm in the relationship between artificial intelligence and the economy. While the science of artificial intelligence made slow but steady progress for decades, only recently did progress rapidly accelerate, allowing these academic achievements to be translated into real-world use-cases.

The technical challenges of beating a human at the game of Go were already familiar to me. As a young Ph.D. student researching artificial intelligence at Carnegie Mellon University, I studied under pioneering AI researcher Raj Reddy. In 1986, I created the first software program to defeat a member of the world championship team for the game Othello, a simplified version of Go played on an eight-by-eight square board. It was quite an accomplishment at the time, but the technology behind it wasn't ready to tackle anything but straightforward board games.

The same held true when IBM's Deep Blue defeated world chess champion Garry Kasparov in a 1997 match dubbed "The Brain's Last Stand." That event had spawned anxiety about when our robot overlords would launch their conquest of humankind, but other than boosting IBM's stock price, the match had no meaningful impact on life in the real world. Artificial intelligence still had few practical applications, and researchers had gone decades without making a truly fundamental breakthrough.

Deep Blue had essentially "brute forced" its way to victory — relying largely on hardware customized to rapidly generate and evaluate positions from each move. It had also required real-life chess champions to add guiding heuristics to the software. Yes, the win was an impressive feat of engineering, but it was based on long-established technology that worked only on very constrained sets of issues. Remove Deep Blue from the geometric simplicity of an eight-by-eight-square chessboard and it wouldn't seem very intelligent at all. In the end, the only job it was threatening to take was that of the world chess champion.

This time, things are different. The Ke Jie versus AlphaGo match was played within the constraints of a Go board, but it is intimately tied up with dramatic changes in the real world. Those changes include the Chinese AI frenzy that AlphaGo's matches sparked amid the underlying technology that powered it to victory.

AlphaGo runs on deep learning, a groundbreaking approach to artificial intelligence that has turbocharged the cognitive capabilities of machines. Deep-learning-based programs can now do a better job than humans at identifying faces, recognizing speech, and issuing loans. For decades, the artificial intelligence revolution always looked to be five years away. But with the development of deep learning over the past few years, that revolution has finally arrived. It will usher in an era of massive productivity increases but also widespread disruptions in labor markets — and profound sociopsychological effects on people — as artificial intelligence takes over human jobs across all sorts of industries.

During the Ke Jie match, it wasn't the AI-driven killer robots some prominent technologists warn of that frightened me. It was the real-world demons that could be conjured up by mass unemployment and the resulting social turmoil. The threat to jobs is coming far faster than most experts anticipated, and it will not discriminate by the color of one's collar, instead striking the highly trained and poorly educated alike. On the day of that remarkable match between AlphaGo and Ke Jie, deep learning was dethroning humankind's best Go player. That same job-eating technology is coming soon to a factory and an office near you.

But in that same match, I also saw a reason for hope. Two hours and fifty-one minutes into the match, Ke Jie had hit a wall. He'd given all that he could to this game, but he knew it wasn't going to be enough. Hunched low over the board, he pursed his lips and his eyebrow began to twitch. Realizing he couldn't hold his emotions in any longer, he removed his glasses and used the back of his hand to wipe tears from both of his eyes. It happened in a flash, but the emotion behind it was visible for all to see.

Those tears triggered an outpouring of sympathy and support for Ke. Over the course of these three matches, Ke had gone on a roller-coaster of human emotion: confidence, anxiety, fear, hope, and heartbreak. It had showcased his competitive spirit, but I saw in those games an act of genuine love: a willingness to tangle with an unbeatable opponent out of pure love for the game, its history, and the people who play it. Those people who watched Ke's frustration responded in kind. AlphaGo may have been the winner, but Ke became the people's champion. In that connection — human beings giving and receiving love — I caught a glimpse of how humans will find work and meaning in the age of artificial intelligence.

I believe that the skillful application of AI will be China's greatest opportunity to catch up with — and possibly surpass — the United States. But more important, this shift will create an opportunity for all people to rediscover what it is that makes us human.

To understand why, we must first grasp the basics of the technology and how it is set to transform our world.

A BRIEF HISTORY OF DEEP LEARNING

Machine learning — the umbrella term for the field that includes deep learning — is a history-altering technology but one that is lucky to have survived a tumultuous half-century of research. Ever since its inception, artificial intelligence has undergone a number of boom-and-bust cycles. Periods of great promise have been followed by "AI winters," when a disappointing lack of practical results led to major cuts in funding. Understanding what makes the arrival of deep learning different requires a quick recap of how we got here.

Back in the mid-1950s, the pioneers of artificial intelligence set themselves an impossibly lofty but well-defined mission: to recreate human intelligence in a machine. That striking combination of the clarity of the goal and the complexity of the task would draw in some of the greatest minds in the emerging field of computer science: Marvin Minsky, John McCarthy, and Herbert Simon.

As a wide-eyed computer science undergrad at Columbia University in the early 1980s, all of this seized my imagination. I was born in Taiwan in the early 1960s but moved to Tennessee at the age of eleven and finished middle and high school there. After four years at Columbia in New York, I knew that I wanted to dig deeper into AI. When applying for computer science Ph.D. programs in 1983, I even wrote this somewhat grandiose description of the field in my statement of purpose: "Artificial intelligence is the elucidation of the human learning process, the quantification of the human thinking process, the explication of human behavior, and the understanding of what makes intelligence possible. It is men's final step to understand themselves, and I hope to take part in this new, but promising science."

That essay helped me get into the top-ranked computer science department of Carnegie Mellon University, a hotbed for cutting-edge AI research. It also displayed my naiveté about the field, both overestimating our power to understand ourselves and underestimating the power of AI to produce superhuman intelligence in narrow spheres.

By the time I began my Ph.D., the field of artificial intelligence had forked into two camps: the "rule-based" approach and the "neural networks" approach. Researchers in the rule-based camp (also sometimes called "symbolic systems" or "expert systems") attempted to teach computers to think by encoding a series of logical rules: If X, then Y. This approach worked well for simple and well-defined games ("toy problems") but fell apart when the universe of possible choices or moves expanded. To make the software more applicable to real-world problems, the rule-based camp tried interviewing experts in the problems being tackled and then coding their wisdom into the program's decision-making (hence the "expert systems" moniker).

The "neural networks" camp, however, took a different approach. Instead of trying to teach the computer the rules that had been mastered by a human brain, these practitioners tried to reconstruct the human brain itself. Given that the tangled webs of neurons in animal brains were the only thing capable of intelligence as we knew it, these researchers figured they'd go straight to the source. This approach mimics the brain's underlying architecture, constructing layers of artificial neurons that can receive and transmit information in a structure akin to our networks of biological neurons. Unlike the rule-based approach, builders of neural networks generally do not give the networks rules to follow in making decisions. They simply feed lots and lots of examples of a given phenomenon — pictures, chess games, sounds — into the neural networks and let the networks themselves identify patterns within the data. In other words, the less human interference, the better.

Differences between the two approaches can be seen in how they might approach a simple problem, identifying whether there is a cat in a picture. The rule-based approach would attempt to lay down "if-then" rules to help the program make a decision: "If there are two triangular shapes on top of a circular shape, then there is probably a cat in the picture." The neural network approach would instead feed the program millions of sample photos labeled "cat" or "no cat," letting the program figure out for itself what features in the millions of images were most closely correlated to the "cat" label.

During the 1950s and 1960s, early versions of artificial neural networks yielded promising results and plenty of hype. But then in 1969, researchers from the rule-based camp pushed back, convincing many in the field that neural networks were unreliable and limited in their use. The neural networks approach quickly went out of fashion, and AI plunged into one of its first "winters" during the 1970s.

Over the subsequent decades, neural networks enjoyed brief stints of prominence, followed by near-total abandonment. In 1988, I used a technique akin to neural networks (Hidden Markov Models) to create Sphinx, the world's first speaker-independent program for recognizing continuous speech. That achievement landed me a profile in the *New York Times*. But it wasn't enough to save neural networks from once again falling out of favor, as AI reentered a prolonged ice age for most of the 1990s.

What ultimately resuscitated the field of neural networks — and sparked the AI renaissance we are living through today — were changes to two of the key raw ingredients that neural networks feed on, along with one major technical breakthrough. Neural networks require large amounts of two things: computing power and data. The data "trains" the program to recognize patterns by giving it many examples, and the computing power lets the program parse those examples at high speeds.

Both data and computing power were in short supply at the dawn of the field in the 1950s. But in the intervening decades, all that has changed. Today, your smartphone holds millions of times more processing power than the leading cutting-edge computers that NASA used to send Neil Armstrong to the moon in 1969. And the internet has led to an explosion of all kinds of digital data: text, images, videos, clicks, purchases, Tweets, and so on. Taken together, all of this has given researchers copious amounts of rich data on which to train their networks, as well as plenty of cheap computing power for that training.

But the networks themselves were still severely limited in what they could do. Accurate results to complex problems required many layers of artificial neurons, but researchers hadn't found a way to efficiently train those layers as they were added. Deep learning's big technical break finally arrived in the mid-2000s, when leading researcher Geoffrey Hinton discovered a way to efficiently train those new layers in neural networks. The result was like giving steroids to the old neural networks, multiplying their power to perform tasks such as speech and object recognition.

Soon, these juiced-up neural networks — now rebranded as "deep learning" — could outperform older models at a variety of tasks. But years of ingrained prejudice against the neural networks approach led many AI researchers to overlook this "fringe" group that claimed outstanding results. The turning point came in 2012, when a neural network built by Hinton's team demolished the competition in an international computer vision contest.

After decades spent on the margins of AI research, neural networks hit the mainstream overnight, this time in the form of deep learning. That breakthrough promised to thaw the ice from the latest AI winter, and for the first time truly bring AI's power to bear on a range of real-world problems. Researchers, futurists, and tech CEOs all began buzzing about the massive potential of the field to decipher human speech, translate documents, recognize images, predict consumer behavior, identify fraud, make lending decisions, help robots "see," and even drive a car.

PULLING BACK THE CURTAIN ON DEEP LEARNING

So how does deep learning do this? Fundamentally, these algorithms use massive amounts of data from a specific domain to make a decision that optimizes for a desired outcome. It does this by training itself to recognize deeply buried patterns and correlations connecting the many data points to the desired outcome. This pattern-finding process is easier when the data is labeled with that desired outcome — "cat" versus "no cat"; "clicked" versus "didn't click"; "won game" versus "lost game." It can then draw on its extensive knowledge of these correlations — many of which are invisible or irrelevant to human observers — to make better decisions than a human could.

Doing this requires massive amounts of relevant data, a strong algorithm, a narrow domain, and a concrete goal. If you're short any one of these, things fall apart. Too little data? The algorithm doesn't have enough examples to uncover meaningful correlations. Too broad a goal? The algorithm lacks clear benchmarks to shoot for in optimization.

Deep learning is what's known as "narrow AI" — intelligence that takes data from one specific domain and applies it to optimizing one specific outcome. While impressive, it is still a far cry from "general AI," the all-purpose technology that can do everything a human can.

Deep learning's most natural application is in fields like insurance and making loans. Relevant data on borrowers is abundant (credit score, income, recent credit-card usage), and the goal to optimize for is clear (minimize default rates). Taken one step further, deep learning will power self-driving cars by helping them to "see" the world around them — recognize patterns in the camera's pixels (red octagons), figure out what they correlate to (stop signs), and use that information to make decisions (apply pressure to the brake to slowly stop) that optimize for your desired outcome (deliver me safely home in minimal time).

People are so excited about deep learning precisely because its core power — its ability to recognize a pattern, optimize for a specific outcome, make a decision — can be applied to so many different kinds of everyday problems. That's why companies like Google and Facebook have scrambled to snap up the small core of deep-learning experts, paying them millions of dollars to pursue ambitious research projects. In 2013, Google acquired the startup founded by Geoffrey Hinton, and the following year scooped up British AI startup DeepMind — the company that went on to build AlphaGo — for over $500 million. The results of these projects have continued to awe observers and grab headlines. They've shifted the cultural zeitgeist and given us a sense that we stand at the precipice of a new era, one in which machines will radically empower and/or violently displace human beings.

AI AND INTERNATIONAL RESEARCH

But where was China in all this? The truth is, the story of the birth of deep learning took place almost entirely in the United States, Canada, and the United Kingdom. After that, a smaller number of Chinese entrepreneurs and venture-capital funds like my own began to invest in this area. But the great majority of China's technology community didn't properly wake up to the deep-learning revolution until its Sputnik Moment in 2016, a full decade behind the field's breakthrough academic paper and four years after it proved itself in the computer vision competition.

American universities and technology companies have for decades reaped the rewards of the country's ability to attract and absorb talent from around the globe. Progress in AI appeared to be no different. The United States looked to be out to a commanding lead, one that would only grow as these elite researchers leveraged Silicon Valley's generous funding environment, unique culture, and powerhouse companies. In the eyes of most analysts, China's technology industry was destined to play the same role in global AI that it had for decades: that of the copycat who lagged far behind the cutting edge.

As I demonstrate in the following chapters, that analysis is wrong. It is based on outdated assumptions about the Chinese technology environment, as well as a more fundamental misunderstanding of what is driving the ongoing AI revolution. The West may have sparked the fire of deep learning, but China will be the biggest beneficiary of the heat the AI fire is generating. That global shift is the product of two transitions: from the age of discovery to the age of implementation, and from the age of expertise to the age of data.

Core to the mistaken belief that the United States holds a major edge in AI is the impression that we are living in an age of discovery, a time in which elite AI researchers are constantly breaking down old paradigms and finally cracking longstanding mysteries. This impression has been fed by a constant stream of breathless media reports announcing the latest feat performed by AI: diagnosing certain cancers better than doctors, beating human champions at the bluff-heavy game of Texas Hold'em, teaching itself how to master new skills with zero human interference. Given this flood of media attention to each new achievement, the casual observer—or even expert analyst—would be forgiven for believing that we are consistently breaking fundamentally new ground in artificial intelligence research.

I believe this impression is misleading. Many of these new milestones are, rather, merely the application of the past decade's breakthroughs—primarily deep learning but also complementary technologies like reinforcement learning and transfer learning—to new problems. What these researchers are doing requires great skill and deep knowledge: the ability to tweak complex mathematical algorithms, to manipulate massive amounts of data, to adapt neural networks to different problems. That often takes Ph.D.-level expertise in these fields. But these advances are incremental improvements and optimizations that leverage the dramatic leap forward of deep learning.

THE AGE OF IMPLEMENTATION

What they really represent is the application of deep learning's incredible powers of pattern recognition and prediction to different spheres, such as diagnosing a disease, issuing an insurance policy, driving a car, or translating a Chinese sentence into readable English. They do *not* signify rapid progress toward "general AI" or any other similar breakthrough on the level of deep learning. This is the age of implementation, and the companies that cash in on this time period will need talented entrepreneurs, engineers, and product managers.

Deep-learning pioneer Andrew Ng has compared AI to Thomas Edison's harnessing of electricity: a breakthrough technology on its own, and one that once harnessed can be applied to revolutionizing dozens of different industries. Just as nineteenth-century entrepreneurs soon began applying the electricity breakthrough to cooking food, lighting rooms, and powering industrial equipment, today's AI entrepreneurs are doing the same with deep learning. Much of the difficult but abstract work of AI research has been done, and it's now time for entrepreneurs to roll up their sleeves and get down to the dirty work of turning algorithms into sustainable businesses.

That in no way diminishes the current excitement around AI; implementation is what makes academic advances meaningful and what will truly end up changing the fabric of our daily lives. The age of implementation means we will finally see real-world applications after decades of promising research, something I've been looking forward to for much of my adult life.

But making that distinction between discovery and implementation is core to understanding how AI will shape our lives and what — or which country — will primarily drive that progress. During the age of discovery, progress was driven by a handful of elite thinkers, virtually all of whom were clustered in the United States and Canada. Their research insights and unique intellectual innovations led to a sudden and monumental ramping up of what computers can do. Since the dawn of deep learning, no other group of researchers or engineers has come up with innovation on that scale.

THE AGE OF DATA

This brings us to the second major transition, from the age of expertise to the age of data. Today, successful AI algorithms need three things: big data, computing power, and the work of strong—but not necessarily elite—AI algorithm engineers. Bringing the power of deep learning to bear on new problems requires all three, but in this age of implementation, data is the core. That's because once computing power and engineering talent reach a certain threshold, the quantity of data becomes decisive in determining the overall power and accuracy of an algorithm.

In deep learning, there's no data like more data. The more examples of a given phenomenon a network is exposed to, the more accurately it can pick out patterns and identify things in the real world. Given much more data, an algorithm designed by a handful of mid-level AI engineers usually outperforms one designed by a world-class deep-learning researcher. Having a monopoly on the best and the brightest just isn't what it used to be.

Elite AI researchers still have the potential to push the field to the next level, but those advances have occurred once every several decades. While we wait for the next breakthrough, the burgeoning availability of data will be the driving force behind deep learning's disruption of countless industries around the world.

ADVANTAGE CHINA

Realizing the newfound promise of electrification a century ago required four key inputs: fossil fuels to generate it, entrepreneurs to build new businesses around it, electrical engineers to manipulate it, and a supportive government to develop the underlying public infrastructure. Harnessing the power of AI today—the "electricity" of the twenty-first century—requires four analogous inputs: abundant data, hungry entrepreneurs, AI scientists, and an AI-friendly policy environment. By looking at the relative strengths of China and the United States in these four categories, we can predict the emerging balance of power in the AI world order.

Both of the transitions described on the previous pages — from discovery to implementation, and from expertise to data — now tilt the playing field toward China. They do this by minimizing China's weaknesses and amplifying its strengths. Moving from discovery to implementation reduces one of China's greatest weak points (outside-the-box approaches to research questions) and also leverages the country's most significant strength: scrappy entrepreneurs with sharp instincts for building robust businesses. The transition from expertise to data has a similar benefit, downplaying the importance of the globally elite researchers that China lacks and maximizing the value of another key resource that China has in abundance, data.

Silicon Valley's entrepreneurs have earned a reputation as some of the hardest working in America, passionate young founders who pull all-nighters in a mad dash to get a product out, and then obsessively iterate that product while seeking out the next big thing. Entrepreneurs there do indeed work hard. But I've spent decades deeply embedded in both Silicon Valley and China's tech scene, working at Apple, Microsoft, and Google before incubating and investing in dozens of Chinese startups. I can tell you that Silicon Valley looks downright sluggish compared to its competitor across the Pacific.

China's successful internet entrepreneurs have risen to where they are by conquering the most cutthroat competitive environment on the planet. They live in a world where speed is essential, copying is an accepted practice, and competitors will stop at nothing to win a new market. Every day spent in China's startup scene is a trial by fire, like a day spent as a gladiator in the Coliseum. The battles are life or death, and your opponents have no scruples.

The only way to survive this battle is to constantly improve one's product but also to innovate on your business model and build a "moat" around your company. If one's only edge is a single novel idea, that idea will invariably be copied, your key employees will be poached, and you'll be driven out of business by VC-subsidized competitors. This rough-and-tumble environment makes a strong contrast to Silicon Valley, where copying is stigmatized and many companies are allowed to coast on the basis of one original idea or lucky break. That lack of competition can lead to a certain level of complacency, with entrepreneurs failing to explore all the possible iterations of their first innovation. The messy markets and dirty tricks of China's "copycat" era produced some questionable companies, but they also incubated a generation of the world's most nimble, savvy, and nose-to-the-grindstone entrepreneurs. These entrepreneurs will be the secret sauce that helps China become the first country to cash in on AI's age of implementation.

These entrepreneurs will have access to the other "natural resource" of China's tech world: an overabundance of data. China has already surpassed the United States in terms of sheer volume as the number one producer of data. That data is not just impressive in quantity, but thanks to China's unique technology ecosystem — an alternate universe of products and functions not seen anywhere else — that data is tailor-made for building profitable AI companies.

Until about five years ago, it made sense to directly compare the progress of Chinese and U.S. internet companies as one would describe a race. They were on roughly parallel tracks, and the United States was slightly ahead of China. But around 2013, China's internet took a right turn. Rather than following in the footsteps or outright copying of American companies, Chinese entrepreneurs began developing products and services with simply no analog in Silicon Valley. Analysts describing China used to invoke simple Silicon Valley–based analogies when describing Chinese companies — "the Facebook of China," "the Twitter of China" — but in the last few years, in many cases these labels stopped making sense. The Chinese internet had morphed into an alternate universe.

Chinese urbanites began paying for real-world purchases with bar codes on their phones, part of a mobile payments revolution unseen anywhere else. Armies of food deliverymen and on-demand masseuses riding electric scooters clogged the streets of Chinese cities. They represented a tidal wave of online-to-offline (O2O) startups that brought the convenience of e-commerce to bear on real-world services like restaurant food or manicures. Soon after that came the millions of brightly colored shared bikes that users could pick up or lock up anywhere just by scanning a bar code with their phones.

Tying all these services together was the rise of China's super-app, WeChat, a kind of digital Swiss Army knife for modern life. We-Chat users began sending text and voice messages to friends, paying for groceries, booking doctors' appointments, filing taxes, unlocking shared bikes, and buying plane tickets, all without ever leaving the app. WeChat became the universal social app, one in which different types of group chats — formed with coworkers and friends or around interests — were used to negotiate business deals, organize birthday parties, or discuss modern art. It brought together a grab-bag of essential functions that are scattered across a dozen apps in the United States and elsewhere.

China's alternate digital universe now creates and captures oceans of new data about the real world. That wealth of information on users — their location every second of the day, how they commute, what foods they like, when and where they buy groceries and beer — will prove invaluable in the era of AI implementation. It gives these companies a detailed treasure trove of these users' daily habits, one that can be combined with deep-learning algorithms to offer tailor-made services ranging from financial auditing to city planning. It also vastly outstrips what Silicon Valley's leading companies can decipher from your searches, "likes," or occasional online purchases. This unparalleled trove of real-world data will give Chinese companies a major leg up in developing AI-driven services.

THE HAND ON THE SCALES

These recent and powerful developments naturally tilt the balance of power in China's direction. But on top of this natural rebalancing, China's government is also doing everything it can to tip the scales. The Chinese government's sweeping plan for becoming an AI super-power pledged widespread support and funding for AI research, but most of all it acted as a beacon to local governments throughout the country to follow suit. Chinese governance structures are more complex than most Americans assume; the central government does not simply issue commands that are instantly implemented throughout the nation. But it does have the ability to pick out certain long-term goals and mobilize epic resources to push in that direction. The country's lightning-paced development of a sprawling high-speed rail network serves as a living example.

Local government leaders responded to the AI surge as though they had just heard the starting pistol for a race, fully competing with each other to lure AI companies and entrepreneurs to their regions with generous promises of subsidies and preferential policies. That race is just getting started, and exactly how much impact it will have on China's AI development is still unclear. But whatever the outcome, it stands in sharp contrast to a U.S. government that deliberately takes a hands-off approach to entrepreneurship and is actively slashing funding for basic research.

Putting all these pieces together — the dual transitions into the age of implementation and the age of data, China's world-class entrepreneurs and proactive government — I believe that China will soon match or even overtake the United States in developing and deploying artificial intelligence. In my view, that lead in AI deployment will translate into productivity gains on a scale not seen since the Industrial Revolution. PricewaterhouseCoopers estimates AI deployment will add $15.7 trillion to global GDP by 2030. China is predicted to take home $7 trillion of that total, nearly double North America's $3.7 trillion in gains. As the economic balance of power tilts in China's favor, so too will political influence and "soft power," the country's cultural and ideological footprint around the globe.

This new AI world order will be particularly jolting to Americans who have grown accustomed to a near-total dominance of the technological sphere. For as far back as many of us can remember, it was American technology companies that were pushing their products and their values on users around the globe. As a result, American companies, citizens, and politicians have forgotten what it feels like to be on the receiving end of these exchanges, a process that often feels akin to "technological colonization." China does not intend to use its advantage in the AI era as a platform for such colonization, but AI-induced disruptions to the political and economic order will lead to a major shift in how all countries experience the phenomenon of digital globalization.

THE REAL CRISES

Significant as this jockeying between the world's two superpowers will be, it pales in comparison to the problems of job losses and growing inequality — both domestically and between countries — that AI will conjure. As deep learning washes over the global economy, it will indeed wipe out billions of jobs up and down the economic ladder: accountants, assembly line workers, warehouse operators, stock analysts, quality control inspectors, truckers, paralegals, and even radiologists, just to name a few.

Human civilization has in the past absorbed similar technology-driven shocks to the economy, turning hundreds of millions of farmers into factory workers over the nineteenth and twentieth centuries. But none of these changes ever arrived as quickly as AI. Based on the current trends in technology advancement and adoption, I predict that within fifteen years, artificial intelligence will technically be able to replace around 40 to 50 percent of jobs in the United States. Actual job losses may end up lagging those technical capabilities by an additional decade, but I forecast that the disruption to job markets will be very real, very large, and coming soon.

Rising in tandem with unemployment will be astronomical wealth in the hands of the new AI tycoons. Uber is already one of the most valuable startups in the world, even while giving around 75 percent of the money earned from each ride to the driver. To that end, how valuable would Uber become if in the span of a couple of years, the company was able to replace every single human driver with an AI-powered self-driving car? Or if banks could replace all their mortgage lenders with algorithms that issued smarter loans with much lower default rates — all without human interference? Similar transformations will soon play out across industries like trucking, insurance, manufacturing, and retail.

Further concentrating those profits is the fact that AI naturally trends toward winner-take-all economics within an industry. Deep learning's relationship with data fosters a virtuous circle for strengthening the best products and companies: more data leads to better products, which in turn attract more users, who generate more data that further improves the product. That combination of data and cash also attracts the top AI talent to the top companies, widening the gap between industry leaders and laggards.

In the past, the dominance of physical goods and limits of geography helped rein in consumer monopolies. (U.S. antitrust laws didn't hurt either.) But going forward, digital goods and services will continue eating up larger shares of the consumer pie, and autonomous trucks and drones will dramatically slash the cost of shipping physical goods. Instead of a dispersion of industry profits across different companies and regions, we will begin to see greater and greater concentration of these astronomical sums in the hands of a few, all while unemployment lines grow longer.

THE AI WORLD ORDER

Inequality will not be contained within national borders. China and the United States have already jumped out to an enormous lead over all other countries in artificial intelligence, setting the stage for a new kind of bipolar world order. Several other countries — the United Kingdom, France, and Canada, to name a few — have strong AI research labs staffed with great talent, but they lack the venture-capital ecosystem and large user bases to generate the data that will be key to the age of implementation. As AI companies in the United States and China accumulate more data and talent, the virtuous cycle of data-driven improvements is widening their lead to a point where it will become insurmountable. China and the United States are currently incubating the AI giants that will dominate global markets and extract wealth from consumers around the globe.

At the same time, AI-driven automation in factories will undercut the one economic advantage developing countries historically possessed: cheap labor. Robot-operated factories will likely relocate to be closer to their customers in large markets, pulling away the ladder that developing countries like China and the "Asian Tigers" of South Korea and Singapore climbed up on their way to becoming high-income, technology-driven economies. The gap between the global haves and have-nots will widen, with no known path toward closing it.

The AI world order will combine winner-take-all economics with an unprecedented concentration of wealth in the hands of a few companies in China and the United States. This, I believe, is the real underlying threat posed by artificial intelligence: tremendous social disorder and political collapse stemming from widespread unemployment and gaping inequality.

Tumult in job markets and turmoil across societies will occur against the backdrop of a far more personal and human crisis —a psychological loss of one's purpose. For centuries, human beings have filled their days by working: trading their time and sweat for shelter and food. We've built deeply entrenched cultural values around this exchange, and many of us have been conditioned to derive our sense of self-worth from the act of daily work. The rise of artificial intelligence will challenge these values and threatens to undercut that sense of life-purpose in a vanishingly short window of time.

These challenges are momentous but not insurmountable. In recent years, I myself faced a mortal threat and a crisis of purpose in my own personal life. That experience transformed me and opened my eyes to potential solutions to the AI-induced jobs crisis I foresee. Tackling these problems will require a combination of clear-eyed analysis and profound philosophical examination of what matters in our lives, a task for both our minds and our hearts. In the closing chapters of this book I outline my own vision for a world in which humans not only coexist alongside AI but thrive with it.

Getting ourselves there—on a technological, social, and human level—requires that we first understand how we arrived here. To do that we must look back fifteen years to a time when China was derided as a land of copycat companies and Silicon Valley stood proud and alone on the technological cutting edge.

2

COPYCATS IN THE COLISEUM

They called him The Cloner. Wang Xing (pronounced "Wang Shing") made his mark on the early Chinese internet as a serial copycat, a bizarre mirror image of the revered serial entrepreneurs of Silicon Valley. In 2003, 2005, 2007, and again in 2010, Wang took America's hottest startup of the year and copied it for Chinese users.

It all began when he stumbled on the pioneering social network Friendster while pursuing an engineering Ph.D. at the University of Delaware. The concept of a virtual network of friendships instantly clicked with Wang's background in computer networking, and he dropped out of his doctoral program to return to China to recreate Friendster. On this first project, he chose not to clone Friendster's exact design. Rather, he and a couple of friends just took the core concept of the digital social network and built their own user interface around it. The result was, in Wang's words, "ugly," and the site failed to take off.

Two years later, Facebook was storming college campuses with its clean design and niche targeting of students. Wang adopted both when he created Xiaonei ("On Campus"). The network was exclusive to Chinese college students, and the user interface was an exact copy of Mark Zuckerberg's site. Wang meticulously recreated the home page, profiles, tool bars, and color schemes of the Palo Alto startup. Chinese media reported that the earliest version of Xiaonei even went so far as to put Facebook's own tagline, "A Mark Zuckerberg Production," at the bottom of each page.

Xiaonei was a hit, but one that Wang sold off too early. As the site grew rapidly, he couldn't raise enough money to pay for server costs and was forced to accept a buyout. Under new ownership, a rebranded version of Xiaonei — now called Renren, "Everybody" — eventually raised $740 million during its 2011 debut on the New York Stock Exchange. In 2007, Wang was back at it again, making a precise copy of the newly founded Twitter. The clone was done so well that if you changed the language and the URL, users could easily be fooled into thinking they were on the original Twitter. The Chinese site, Fanfou, thrived for a moment but was soon shut down over politically sensitive content. Then, three years later Wang took the business model of red-hot Groupon and turned it into the Chinese group-buying site Meituan.

To the Silicon Valley elite, Wang was shameless. In the mythology of the valley, few things are more stigmatized than blindly aping the establishment. It was precisely this kind of copycat entrepreneurship that would hold China back, or so the conventional wisdom said, and would prevent China from building truly innovative technology companies that could "change the world."

Even some entrepreneurs in China felt that Wang's pixel-for-pixel cloning of Facebook and Twitter went too far. Yes, Chinese companies often imitated their American peers, but you could at least localize or add a touch of your own style. But Wang made no apologies for his mimic sites. Copying was a piece of the puzzle, he said, but so was his choice of which sites to copy and his execution on the technical and business fronts.

In the end, it was Wang who would get the last laugh. By late 2017, Groupon's market cap had shriveled to $2.58 billion, with its stock trading at under one-fifth the price of its 2011 initial public offering (IPO). The former darling of the American startup world had been stagnant for years and slow to react when the group-buying craze faded. Meanwhile, Wang Xing's Meituan had triumphed in a brutally competitive environment, beating out thousands of similar group-buying websites to dominate the field. It then branched out into dozens of new lines of business. It is now the fourth most valuable startup in the world, valued at $30 billion, and

Wang sees Alibaba and Amazon as his main competitors going forward.

In analyzing Wang's success, Western observers make a fundamental mistake. They believe Meituan triumphed by taking a great American idea and simply copying it in the sheltered Chinese internet, a safe space where weak local companies can survive under far less intense competition. This kind of analysis, however, is the result of a deep misunderstanding of the dynamics at play in the Chinese market, and it reveals an egocentrism that defines all internet innovation in relation to Silicon Valley.

In creating his early clones of Facebook and Twitter, Wang was in fact relying entirely on the Silicon Valley playbook. This first phase of the copycat era — Chinese startups cloning Silicon Valley websites — helped build up baseline engineering and digital entrepreneurship skills that were totally absent in China at the time. But it was a second phase — Chinese startups taking inspiration from an American business model and then fiercely competing against each other to adapt and optimize that model specifically for Chinese users — that turned Wang Xing into a world-class entrepreneur.

Wang didn't build a $30 billion company by simply bringing the group-buying business model to China. Over five thousand companies did the exact same thing, including Groupon itself. The American company even gave itself a major leg up on local copycats by partnering with a leading Chinese internet portal. Between 2010 and 2013, Groupon and its local impersonators waged an all-out war for market share and customer loyalty, burning billions of dollars and stopping at nothing to slay the competition.

The battle royal for China's group-buying market was a microcosm of what China's internet ecosystem had become: a coliseum where hundreds of copycat gladiators fought to the death. Amid the chaos and bloodshed, the foreign first-movers often proved irrelevant. It was the domestic combatants who pushed each other to be faster, nimbler, leaner, and meaner. They aggressively copied each other's product innovations, cut prices to the bone, launched smear campaigns, forcibly deinstalled competing software, and even reported rival CEOs to the police. For these gladiators, no dirty trick or underhanded maneuver was out of bounds. They deployed tactics that would make Uber founder Travis Kalanick blush. They also demonstrated a fanatical around-the-clock work ethic that would send Google employees running to their nap pods.

Silicon Valley may have found the copying undignified and the tactics unsavory. In many cases, it was. But it was precisely this widespread cloning — the onslaught of thousands of mimicking competitors — that forced companies to innovate. Survival in the internet coliseum required relentlessly iterating products, controlling costs, executing flawlessly, generating positive PR, raising money at exaggerated valuations, and seeking ways to build a robust business "moat" to keep the copycats out. Pure copycats never made for great companies, and they couldn't survive inside this coliseum. But the trial-by-fire competitive landscape created when one is surrounded by ruthless copycats had the result of forging a generation of the most tenacious entrepreneurs on earth.

As we enter the age of AI implementation, this cutthroat entrepreneurial environment will be one of China's core assets in building a machine-learning-driven economy. The dramatic transformation that deep learning promises to bring to the global economy won't be delivered by isolated researchers producing novel academic results in the elite computer science labs of MIT or Stanford. Instead, it will be delivered by down-to-earth, profit-hungry entrepreneurs teaming up with AI experts to bring the transformative power of deep learning to bear on real-world industries.

Over the coming decade, China's gladiator entrepreneurs will fan out across hundreds of industries, applying deep learning to any problem that shows the potential for profit. If artificial intelligence is the new electricity, Chinese entrepreneurs will be the tycoons and tinkerers who electrify everything from household appliances to homeowners' insurance. Their knack for endlessly tweaking business models and sniffing out profits will yield an incredible array of practical — maybe even life-changing — applications. These will be deployed in their home country and then pushed abroad, potentially taking over most developing markets around the globe.

Corporate America is unprepared for this global wave of Chinese entrepreneurship because it fundamentally misunderstood the secret to The Cloner's success. Wang Xing didn't succeed because he'd been a copycat. He triumphed because he'd become a gladiator.

CONTRASTING CULTURES

Startups and the entrepreneurs who found them are not born in a vacuum. Their business models, products, and core values constitute an expression of the unique cultural time and place in which they come of age.

Silicon Valley's and China's internet ecosystems grew out of very different cultural soil. Entrepreneurs in the valley are often the children of successful professionals, such as computer scientists, dentists, engineers, and academics. Growing up they were constantly told that they—yes, *they* in particular—could change the world. Their undergraduate years were spent learning the art of coding from the world's leading researchers but also basking in the philosophical debates of a liberal arts education. When they arrived in Silicon Valley, their commutes to and from work took them through the gently curving, tree-lined streets of suburban California.

It's an environment of abundance that lends itself to lofty thinking, to envisioning elegant technical solutions to abstract problems. Throw in the valley's rich history of computer science breakthroughs, and you've set the stage for the geeky-hippie hybrid ideology that has long defined Silicon Valley. Central to that ideology is a wide-eyed techno-optimism, a belief that every person and company can truly change the world through innovative thinking. Copying ideas or product features is frowned upon as a betrayal of the zeitgeist and an act that is beneath the moral code of a true entrepreneur. It's all about "pure" innovation, creating a totally original product that generates what Steve Jobs called a "dent in the universe."

Startups that grow up in this kind of environment tend to be *mission-driven*. They start with a novel idea or idealistic goal, and they build a company around that. Company mission statements are clean and lofty, detached from earthly concerns or financial motivations.

In stark contrast, China's startup culture is the yin to Silicon

Valley's yang: instead of being mission-driven, Chinese companies are first and foremost *market-driven*. Their ultimate goal is to make money, and they're willing to create any product, adopt any model, or go into any business that will accomplish that objective. That mentality leads to incredible flexibility in business models and execution, a perfect distillation of the "lean startup" model often praised in Silicon Valley. It doesn't matter where an idea came from or who came up with it. All that matters is whether you can execute it to make a financial profit. The core motivation for China's market-driven entrepreneurs is not fame, glory, or changing the world. Those things are all nice side benefits, but the grand prize is getting rich, and it doesn't matter how you get there.

Jarring as that mercenary attitude is to many Americans, the Chinese approach has deep historical and cultural roots. Rote memorization formed the core of Chinese education for millennia. Entry into the country's imperial bureaucracy depended on word-for-word memorization of ancient texts and the ability to construct a perfect "eight-legged essay" following rigid stylistic guidelines. While Socrates encouraged his students to seek truth by questioning everything, ancient Chinese philosophers counseled people to follow the rituals of sages from the ancient past. Rigorous copying of perfection was seen as the route to true mastery.

Layered atop this cultural propensity for imitation is the deeply ingrained scarcity mentality of twentieth-century China. Most Chinese tech entrepreneurs are at most one generation away from grinding poverty that stretches back centuries. Many are only children — products of the now-defunct "One Child Policy" — carrying on their backs the expectations of two parents and four grandparents who have invested all their hopes for a better life in this child. Growing up, their parents didn't talk to them about changing the world. Rather, they talked about survival, about a responsibility to earn money so they can take care of their parents when their parents are too old to work in the fields. A college education was seen as the key to escaping generations of grinding poverty, and that required tens of thousands of hours of rote memorization in preparing for China's notoriously competitive entrance exam. During these entrepreneurs' lifetimes, China wrenched itself out of poverty through bold policies and hard work, trading meal tickets for paychecks for equity stakes in startups.

The blistering pace of China's economic rise hasn't alleviated that scarcity mentality. Chinese citizens have watched as industries, cities, and individual fortunes have been created and lost overnight in a Wild West environment where regulations struggled to keep pace with cutthroat market competition. Deng Xiaoping, the Chinese leader who pushed China from Mao-era egalitarianism to market-driven competition, once said that China needed to "let some people get rich first" in order to develop. But the lightning speed of that development only heightened fears and concerns that if you don't move quickly — if you don't grab onto this new trend or jump into that new market — you'll stay poor while others around you get rich.

Combine these three currents — a cultural acceptance of copying, a scarcity mentality, and the willingness to dive into any promising new industry — and you have the psychological foundations of China's internet ecosystem.

This is not meant to preach a gospel of cultural determinism. As someone who has moved between these two countries and cultures, I know that birthplace and heritage are not the sole determinants of behavior. Personal eccentricities and government regulation are hugely important in shaping company behavior. In Beijing, entrepreneurs often joke that Facebook is "the most Chinese company in Silicon Valley" for its willingness to copy from other startups and for Zuckerberg's fiercely competitive streak. Likewise, while working at Microsoft, I saw how government antitrust policy can defang a wolf-like company. But history and culture do matter, and in comparing the evolution of Silicon Valley and Chinese technology, it's crucial to grasp how different cultural melting pots produced different types of companies.

For years, the copycat products that emerged from China's cultural stew were widely mocked by the Silicon Valley elite. They were derided as cheap knockoffs, embarrassments to their creators and unworthy of the attention of true innovators. But those outsiders missed what was brewing beneath the surface. The most valuable product to come out of China's copycat era wasn't a product at all: it was the entrepreneurs themselves.

Twice a day, the Hall of Ancestor Worship comes alive. Located within Beijing's Forbidden City, this was where the emperors of China's last two dynasties once burned incense and performed sacred rituals to honor the Sons of Heaven that came before them. Today, the hall is home to some of the most intricate and ingenious mechanical timepieces ever created. The clock faces themselves convey expert craftsmanship, but it's the impossibly complex mechanical functions embedded in the clocks' structures that draw large crowds for the morning and afternoon performances.

As the seconds tick by, a metal bird darts around a gold cage. Painted wooden lotus flowers open and close their petals, revealing a tiny Buddhist god deep in meditation. A delicately carved elephant lifts its trunk up and down while pulling a miniature carriage in circles. A robotic Chinese figure dressed in the coat of a European scholar uses an ink brush to write out a Chinese aphorism on a miniature scroll, with the robot's own handwriting modeled on the calligraphy of the Chinese emperor who commissioned the piece.

It's a dazzling display, a reminder of the timeless nature of true craftsmanship. Jesuit missionaries brought many of the clocks to China as part of "clock diplomacy," an attempt by Jesuits to charm their way into the imperial court through gifts of advanced European technology. The Qing Dynasty's Qianlong emperor was particularly fond of the clocks, and British manufacturers soon began producing clocks to fit the tastes of the Son of Heaven. Many of the clocks on display at the Hall of Ancestor Worship were the handiwork of Europe's finest artisanal workshops of the seventeenth and eighteenth centuries. These workshops produced an unparalleled combination of artistry, design, and functional engineering. It's a particular alchemy of expertise that feels familiar to many in Silicon Valley today.

While working as the founding president of Google China, I would bring visiting delegations of Google executives here to see the clocks in person. But I didn't do it so they could revel in the genius of their European ancestors. I did it because, on closer inspection, one discovers that many of the finest specimens of European craftsman-ship were created in the southern Chinese city of Guangzhou, which was then called Canton.

After European clocks won the favor of the Chinese emperor, lo-cal workshops sprang up all over China to study and recreate the Western imports. In the southern port cities where Westerners came to trade, China's best craftspeople took apart the ingenious Euro-pean devices, examining each interlocking piece and design flour-ish. They mastered the basics and began producing clocks that were near-exact replicas of the European models. From there, the artisans took the underlying principles of clock-building and began constructing timepieces that embodied Chinese designs and cul-tural traditions: animated Silk Road caravans, lifelike scenes from the streets of Beijing, and the quiet equanimity of Buddhist sutras. These workshops eventually began producing clocks that rivaled or even exceeded the craftsmanship coming out of Europe, all while weaving in an authentically Chinese sensibility.

The Hall of the Ancestors dates back to the Ming Dynasty, and the story of China's own copycat clockmakers played out hundreds of years in the past. But the same cultural currents continue to flow into the present day. As we watched these mechanical marvels twirl and chime, I worried that those currents would soon sweep away the master craftspeople of the twenty-first century who stood all around me.

COPYKITTENS

China's early copycat internet companies looked harmless from the outside, almost cute. During China's first internet boom of the late 1990s, Chinese companies looked to Silicon Valley for talent, funding, and even names for their infant startups. The country's first search engine was the creation of Charles Zhang, a Chinese physicist with a Ph.D. from MIT. While in the United States Zhang had seen the early internet take off, and he wanted to kick-start that same process in his home country. Zhang used investments from his professors at MIT and returned to China, intent on building up the country's core internet infrastructure.

But after a meeting with Yahoo! founder Jerry Yang, Zhang switched his focus to creating a Chinese-language search engine and portal website. He named his new company Sohoo, a not-so-subtle mashup of the Chinese word for "search" (*sou*) and the company's American role model. He soon switched the spelling to "Sohu" to downplay the connection, but this kind of imitation was seen as more flattery than threat to the American web juggernaut. At the time, Silicon Valley saw the Chinese internet as a novelty, an interesting little experiment in a technologically backward country.

Bear in mind that this was an era when copying fueled many parts of the Chinese economy. Factories in the southern part of the country cranked out knockoff luxury bags. Chinese car manufacturers created such close duplicates of foreign models that some dealerships gave customers the option of removing the Chinese company's logo and replacing it with the logo of the more prestigious foreign brand. There was even a knockoff Disneyland, a creepy amusement park on the outskirts of Beijing where employees in replica Mickey and Minnie Mouse suits hugged Chinese children. At the park's entrance hung a sign: "Disneyland is too far, please come to Shijingshan!" While China's enterprising amusement park operators borrowed unabashedly from Disney, Wang Xing was hard at work copying Facebook and then Twitter.

While leading Google China, I experienced firsthand the danger that these clones posed to brand image. Beginning in 2005, I threw myself into building up our Chinese search engine and the trust of Chinese users. But on the evening of December 11, 2008, a major Chinese TV station dedicated a six-minute segment of its national news broadcast to a devastating exposé on Google China. The program showed users searching Google's Chinese site for medical information being served up ads with links to fake medical treatments. The camera zoomed in tight on the computer screen, where Google's Chinese logo hovered ominously above dangerous scams and phony prescription-drug services.

Google China was thrown into a full-on crisis of public trust. After watching the footage, I raced to my computer to conduct the same searches but curiously could not conjure up the results featured on the program. I changed around the words and tweaked my settings but still couldn't navigate to—and then subsequently remove—the offending ads. At the same time, I was immediately flooded with messages from reporters demanding an explanation as to Google China's misleading advertising, but I could only give what probably sounded like a weak excuse: Google works quickly to re-move any problematic advertisements, but the process isn't instan-taneous, and occasionally offending ads may live online for a few hours.

The storm continued to rage on, all while our team kept failing to find or locate the offending ads from the television program. Later that night I received an excited email from one of our engineers. He had figured out why we couldn't reproduce the results: because the search engine shown on the program wasn't Google. It was a Chi-nese copycat search engine that had made a perfect copy of Google—the layout, the fonts, the feel—almost down to the pixel. The site's search results and ads were their own but had been packaged on-line to be indistinguishable from Google China. The engineer had noticed just one tiny difference, a slight variation in the color of one font used. The impersonators had done such a good job that all but one of Google China's seven hundred employees watching onscreen had failed to tell them apart.

The precision copying extended even to the most elegant and cutting-edge hardware. When Steve Jobs launched the original iPhone, he had only a few months' lead time before electronics mar-kets throughout China were selling "mini-iPhones." The fun-size rep-licas looked almost exactly like the real thing but were about half the size and fit squarely in the palm of your hand. They also completely lacked the ability to access the internet via the phone's data plan, making them the dumbest "smartphone" on the market.

American visitors to Beijing would clamor to get their hands on the mini-iPhones, thinking them a great joke gift for friends back home. To those steeped in the innovation mythology of Silicon Val-ley, the mini-iPhones were the perfect metaphor for Chinese tech-nology during the copycat era: a shiny exterior that had been cop-ied from America but a hollow shell that held nothing innovative or even functional. The prevailing American attitude was that people like Wang Xing could copy the look and feel of Facebook, but that the Chinese would never access the mysterious magic of innovation that drove a place like Silicon Valley.

BUILDING BLOCKS AND STUMBLING BLOCKS

Silicon Valley investors take as an article of faith that a pure innovation mentality is the foundation on which companies like Google, Facebook, Amazon, and Apple are built. It was an irrepressible impulse to "think different" that drove people like Steve Jobs, Mark Zuckerberg, and Jeff Bezos to create these companies that would change the world. In that school of thought, China's knockoff clockmakers were headed down a dead-end road. A copycat mentality is a core stumbling block on the path to true innovation. By blindly imitating others — or so the theory goes — you stunt your own imagination and kill the chances of creating an original and innovative product.

But I saw early copycats like Wang Xing's Twitter knockoff not as stumbling blocks but as building blocks. That first act of copying didn't turn into an anti-innovation mentality that its creator could never shake. It was a necessary steppingstone on the way to more original and locally tailored technology products.

The engineering know-how and design sensibility needed to create a world-class technology product don't just appear out of nowhere. In the United States, universities, companies, and engineers have been cultivating and passing down these skillsets for generations. Each generation has its breakout companies or products, but these innovations rest on a foundation of education, mentorship, internships, and inspiration.

China had no such luxury. When Bill Gates founded Microsoft in 1975, China was still in the throes of the Cultural Revolution, a time of massive social upheaval and anti-intellectual fever. When Sergei Brin and Larry Page founded Google in 1998, just 0.2 percent of the Chinese population was connected to the internet, compared with 30 percent in the United States. Early Chinese tech entrepreneurs looking for mentors or model companies within their own country simply couldn't find them. So instead they looked abroad and copied them as best they could.

It was a crude process to be sure, and sometimes an embarrassing one. But it taught these copycats the basics of user interface design, website architecture, and back-end software development. As their clone-like products went live, these market-driven entrepreneurs were forced to grapple with user satisfaction and iterative product development. If they wanted to win the market, they had to beat not just their Silicon Valley inspiration but also droves of similar copycats. They learned what worked and what didn't with Chinese users. They began to iterate, improve, and localize the product to better serve their customers.

And those customers had unique habits and preferences, ways of using software that didn't map neatly onto Silicon Valley's global one-size-fits-all product model. Companies like Google and Facebook are often loath to allow local changes to their core products or business models. They tend to believe in building one thing and building it well. It's an approach that helped them rapidly sweep the globe in the early days of the internet, when most countries lagged so far behind in technology that they couldn't offer any localized alternatives. But as technical know-how has diffused around the globe, it is becoming harder to force people of all countries and cultures into a cookie-cutter mold that was often built in America for Americans.

As a result, when Chinese copycats went head-to-head with their Silicon Valley forefathers, they took that American unwillingness to adapt and weaponized it. Every divergence between Chinese user preferences and a global product became an opening that local competitors could attack. They began tailoring their products and business models to local needs, and driving a wedge between Chinese internet users and Silicon Valley.

"FREE IS NOT A BUSINESS MODEL"

Jack Ma made an art of these kinds of attacks in the early days of the Chinese e-commerce company Alibaba. Ma founded his company in 1999, and for the first couple of years of operation his main competitors were other local Chinese companies. But in 2002, eBay entered the Chinese market. At that time, eBay was the biggest e-commerce company in the world and a darling of both Silicon Valley and Wall Street. Alibaba's online marketplace was derided as another Chinese copycat with no right to be in the same room as the big dogs of Silicon Valley. And so Ma launched a five-year guerrilla war against eBay, turning the foreign company's size against it and relentlessly punishing the invader for failing to adapt to local conditions.

When eBay entered the Chinese market in 2002, they did so by buying the leading Chinese online auction site—not Alibaba but an eBay impersonator called EachNet. The marriage created the ultimate power couple: the top global e-commerce site and China's number one knockoff. eBay proceeded to strip away the Chinese company's user interface, rebuilding the site in eBay's global product image. Company leadership brought in international managers for the new China operations, who directed all traffic through eBay's servers back in the United States. But the new user interface didn't match Chinese web-surfing habits, the new leadership didn't understand Chinese domestic markets, and the trans-Pacific routing of traffic slowed page-loading times. At one point an earthquake under the Pacific Ocean severed key cables and knocked the site offline for a few days.

Meanwhile, Alibaba founder Jack Ma was busy copying eBay's core functions and adapting the business model to Chinese realities. He began by creating an auction-style platform, Taobao, to directly compete with eBay's core business. From there, Ma's team continually tweaked Taobao's functions and tacked on features to meet unique Chinese needs. His strongest localization plays were in payment and revenue models. To overcome a deficit of user trust in online purchases, Ma created Alipay, a payment tool that would hold money from purchases in escrow until the buyer confirmed the receipt of goods. Taobao also added instant messaging functions to allow buyers and sellers to communicate on the platform in real time. These business innovations helped Taobao claw away market share from eBay, whose global product mentality and deep centralization of decision-making power in Silicon Valley made it slow to react and add features.

But Ma's greatest weapon was his deployment of a "freemium" revenue model, the practice of keeping basic functions free while charging for premium services. At the time, eBay charged sellers a fee just to list their products, another fee when the products were sold, and a final fee if eBay-owned PayPal was used for payment. Conventional wisdom held that auction sites or e-commerce marketplace sites needed to do this in order to guarantee steady revenue streams.

But as competition with eBay heated up, Ma developed a new approach: he pledged to make all listings and transactions on Taobao free for the next three years, a promise he soon extended indefinitely. It was an ingenious PR move and a savvy business play. In the short term, it won goodwill from Chinese sellers still leery of internet transactions. Allowing them to list for free helped Ma build a thriving marketplace in a low-trust society. It took years to get there, but in the long term, that marketplace grew so large that in order to get their products noticed, power sellers had to pay Ma for advertisements and higher search rankings. Brands would end up paying even larger premiums to list on Taobao's more high-end sister site, Tmall.

eBay bungled its response. In a condescending press release, the company lectured Ma, claiming "free is not a business model." As a Nasdaq-listed public company, eBay was under pressure to show ever-rising revenues and profits. American public companies tend to treat international markets as cash cows, sources of bonus revenue to which they are entitled by virtue of winning at home. Silicon Valley's richest e-commerce company wasn't about to make an exception to its global model to match the wild pronouncements of a pesky Chinese copycat.

That kind of shortsighted stubbornness sealed eBay's fate in China. Taobao rapidly peeled away users and sellers from the American juggernaut. With eBay's market share in freefall, eBay CEO Meg Whitman briefly relocated to China to try and salvage the operations there. When that didn't work, she invited Ma to Silicon Valley to try and broker a deal. But Ma smelled blood in the water, and he wanted total victory. Within a year, eBay fully retreated from the Chinese market.

THE YELLOW PAGES VERSUS THE BAZAAR

I witnessed this same disconnect between global products and local users while leading Google China. As an extension of perhaps the world's most prestigious internet company, we should have had a major brand advantage. But that linkage back to headquarters in Silicon Valley turned into a big stumbling block when it came to adapting products to wider Chinese audiences. When I launched Google China in 2005, our main competitor was the Chinese search engine Baidu. The website was the creation of Robin Li, a Chinese-born expert in search engines who had experience working in Silicon Valley. Baidu's core functions and minimalist design mimicked Google, but Li relentlessly optimized the site for the search habits of Chinese users.

Those divergent habits were starkest in the ways users interacted with a page of search results. Within focus groups, we were able to track a user's eye movements and clicks across a given page of search results. We used that data to create heat maps of activity on the page: green highlights showed where the user had glanced, yellow highlights where they had stared intently, and red dots marked each of their clicks. Comparing heat maps generated by American and Chinese users makes for a striking contrast.

The American users' maps show a tight clustering of green and yellow in the upper left corner where the top search results appeared, with a couple of red dots for clicks on the top two results. American users remain on the page for around ten seconds before navigating away. In contrast, Chinese users' heat maps look like a hot mess. The upper left corner has the greatest cluster of glances and clicks, but the rest of the page is blanketed in smudges of green and specks of red. Chinese users spent between thirty and sixty seconds on the search page, their eyes darting around almost all the results as they clicked promiscuously.

Eye-tracking maps revealed a deeper truth about the way both sets of users approached search. Americans treated search engines like the Yellow Pages, a tool for simply finding a specific piece of information. Chinese users treated search engines like a shopping mall, a place to check out a variety of goods, try each one on, and eventually pick a few things to buy. For tens of millions of Chinese new to the internet, this was their first exposure to such a variety of information, and they wanted to sample it all.

That strikingly fundamental difference in user attitudes should have led to a number of product modifications for Chinese users. On Google's global search platform, when users clicked on a search result's link, it would navigate them away from the search results page. That meant we were forcing Chinese "shoppers" to pick one item for purchase and then, in effect, kicking them out of the mall. Baidu, by contrast, opened a new browser window for the user for each link clicked. That let users try on various search results without having to "leave the mall."

Given clear evidence of different user needs, I recommended Google make an exception and copy the Baidu model of opening different windows for each click. But the company had a lengthy review process for any changes to core products because those changes "forked" the code and made it more difficult to maintain. Google and other Silicon Valley companies tried hard to avoid that, believing that the elegant products coming out of the Silicon Valley headquarters should be good enough for users around the globe. I fought for months to get this change made and eventually prevailed, but in the meantime Baidu had won over more users with its China-centric product offering.

Battles like this were repeated continuously over my four years with Google. In fairness to Google, headquarters gave us more latitude than most Silicon Valley companies give to their China branches, and we used that leverage to develop many locally optimized features, which won back substantial market share Google had lost in previous years. But headquarters' resistance to forking made each new feature an uphill battle, one that slowed us up and wore us down. Tired of fighting with their own company, many employees left out of frustration.

As a succession of American juggernauts — eBay, Google, Uber, Airbnb, LinkedIn, Amazon — tried and failed to win the Chinese market, Western analysts were quick to chalk up their failures to Chinese government controls. They assumed that the only reason Chinese companies survived was due to government protectionism that hobbled their American opponents.

In my years of experience working for those American companies and now investing in their Chinese competitors, I've found Silicon Valley's approach to China to be a far more important reason for their failure. American companies treat China like just any other market to check off their global list. They don't invest the resources, have the patience, or give their Chinese teams the flexibility needed to compete with China's world-class entrepreneurs. They see the primary job in China as *marketing* their existing products to Chinese users. In reality, they need to put in real work *tailoring* their products for Chinese users or *building* new products from the ground up to meet market demands. Resistance to localization slows down product iteration and makes local teams feel like cogs in a clunky machine.

Silicon Valley companies also lose out on top talent. With so much opportunity now for growth within Chinese startups, the most ambitious young people join or start local companies. They know that if they join the Chinese team of an American company, that company's management will forever see them as "local hires," workers whose utility is limited to their country of birth. They'll never be given a chance to climb the hierarchy at the Silicon Valley headquarters, instead bumping up against the ceiling of a "country manager" for China. The most ambitious young people — the ones who want to make a global impact — chafe at those restrictions, choosing to start their own companies or to climb the ranks at one of China's tech juggernauts. Foreign firms are often left with mild-mannered managers or career salespeople helicoptered in from other countries, people who are more concerned with protecting their salary and stock options than with truly fighting to win the Chinese market. Put those relatively cautious managers up against gladiatorial entrepreneurs who cut their teeth in China's competitive coliseum, and it's always the gladiators who will emerge victorious.

While foreign analysts continued to harp on the question of why American companies couldn't win in China, Chinese companies were busy building better products. Weibo, a micro-blogging platform initially inspired by Twitter, was far faster to expand multimedia functionality and is now worth more than the American company. Didi, the ride-hailing company that duked it out with Uber, dramatically expanded its product offerings and gives more rides each day in China than Uber does across the entire world. Toutiao, a Chinese news platform often likened to BuzzFeed, uses advanced machine-learning algorithms to tailor its content for each user, boosting its valuation many multiples above the American website. Dismissing these companies as copycats relying on government protection in order to succeed blinds analysts to world-class innovation that is happening elsewhere.

But the maturation of China's entrepreneurial ecosystem was about far more than competition with American giants. After companies like Alibaba, Baidu, and Tencent had proven how lucrative China's internet markets could be, new waves of venture capital and talent began to pour into the industry. Markets were heating up, and the number of Chinese startups was growing exponentially. These startups may have taken inspiration from across the ocean, but their real competitors were other domestic companies, and the clashes were taking on all the intensity of a sibling rivalry.

Battles with Silicon Valley may have created some of China's homegrown internet Goliaths, but it was cutthroat Chinese domestic competition that forged a generation of gladiator entrepreneurs.

ALL IS FAIR IN STARTUPS AND WAR

Zhou Hongyi is the kind of guy who likes to pose for pictures with heavy artillery. His 12 million social media followers are regularly treated to pictures of Zhou posing next to cannons or impaling cell phones with a high-powered bow and arrow. For years, one wall of his office was adorned entirely with the shot-up sheets of paper used for handgun target practice. When his PR team submits a stock photo to media outlets, it's sometimes a picture of Zhou dressed in army fatigues, smoke rising in the background and a machine gun leaning by his side.

He is also the fiery founder of some of China's most successful early internet companies. Zhou's first startup sold to Yahoo!, which picked Zhou to head up China operations. Clashing endlessly with the Silicon Valley leadership, Zhou is rumored to have once thrown a chair out an office window during a shouting match. When I led Google China, I would invite Zhou to speak to our leadership team about the unique characteristics of the Chinese market. He took the opportunity to berate the American executives, telling them they were naive and knew nothing about what it took to compete in China. They would, he said, be better off just handing over control to a battle-hardened warrior like him. He later founded China's leading web security software, Qihoo 360 (pronounced "chee-who"), and launched a browser whose logo was an exact copy of Internet Explorer's but done in green.

Zhou embodies the gladiatorial mentality of Chinese internet entrepreneurs. In his world, competition is war and he will stop at nothing to win. In Silicon Valley, his tactics would guarantee social ostracism, antimonopoly investigations, and endless, costly lawsuits. But in the Chinese coliseum, none of these three can hold back combatants. The only recourse when an opponent strikes a low blow is to launch a more damaging counterattack, one that can take the form of copying products, smearing opponents, or even legal detention. Zhou faced all of the above during the "3Q War," a battle between Zhou's Qihoo and QQ, the messaging platform of web juggernaut Tencent.

I witnessed the start of hostilities firsthand one evening in 2010, when Zhou invited me and employees of the newly formed Sinovation Ventures to join his team at a laser tag course outside of Beijing. Zhou was in his element, shooting up the competition, when his cell phone rang. It was an employee with bad news: Tencent had just launched a copycat of Qihoo 360's antivirus product and was automatically installing it on any computer that used QQ. Tencent was already a powerful company that wielded enormous influence through its QQ user base. This was a direct challenge to Qihoo's core business, a matter of corporate life or death in Zhou's mind, as he wrote in his autobiography, *Disruptor*. He immediately called together his team at the laser tag place, and they raced back to their headquarters to formulate a counterattack.

Over the next two months, Zhou pulled out every dirty and desperate trick he could think of to beat back Tencent. Qihoo first created a popular new "privacy protection" software that issued dire safety warnings every time a Tencent product was opened. The warnings were often not based on any real security vulnerability, but it was an effective smear campaign against the stronger company. Qihoo then released a piece of "security" software that could filter all ads within QQ, effectively killing the product's main revenue stream. Soon thereafter, Zhou was on his way to work when he got a phone call: over thirty police officers had raided the Qihoo offices and were waiting there to detain Zhou as part of an investigation. Convinced the raid was orchestrated by Tencent, Zhou drove straight to the airport and fled to Hong Kong to formulate his next move.

Finally, Tencent took the nuclear option: on November 3, 2010, Tencent announced that it would block the use of QQ messaging on any computer that had Qihoo 360, forcing users to choose between the two products. It was the equivalent of Facebook telling users it would block Facebook access for anyone using Google Chrome. The companies were waging total war against each other, with Chinese users' computers as the battleground. Qihoo appealed to users for a three-day "QQ strike," and the government finally stepped in to separate the bloodied combatants. Within a week both QQ and Qihoo 360 had returned to normal functioning, but the scars from these kinds of battles lingered with the entrepreneurs and companies.

Zhou Hongyi was one of the most pugnacious of these entrepreneurs, but dirty tricks and anticompetitive behavior were the norm in the industry. Remember Wang Xing's Facebook copycat, Xiaonei? After he sold it in 2006, the site reemerged as Renren ("Everyone") and became the dominant Facebook-esque social network. But by 2008, Renren faced a scrappy challenger in Kaixin001 (*kaixin* means "happy" in Mandarin). The startup gained traction by initially targeting young urbanites instead of the college students already on

Renren. Kaixinoo1 integrated social networking and gaming with products like "Steal Vegetables," a Farmville knockoff, but one where people were rewarded not for cooperatively farming but for stealing from each other's gardens. The startup quickly became the fastest growing social network around.

Kaixinoo1 was a solid product, but its founder was no gladiator. When he created the network, the URL that he wanted to use —kaixin.com—was already taken, and he didn't want (or possibly couldn't afford) to buy it from its owner. So instead he opted for kaixinoo1.com, which turned out to be a fatal mistake, equivalent to entering the coliseum without a helmet.

The moment Kaixinoo1 became a threat, the owner of Renren simply bought the original www.kaixin.com URL from its owner. He then recreated an exact copy of Kaixinoo1's user interface, changing only the color, and brazenly dubbed it "The Real Kaixin Net." Suddenly, many users trying to sign up for the popular new social network found themselves unwittingly ensnared in Renren's net. Few even knew the difference. Renren later announced it would merge Kaixin.com with Renren, effectively completing its kidnapping of Kaixinoo1 users. The move kneecapped Kaixinoo1's user growth, killed its momentum, and neutralized a major threat to Renren's dominance.

Kaixinoo1 sued its unsavory rival, but the lawsuit couldn't undo the damage from live combat. In April 2011, eighteen months after the lawsuit was filed, a Beijing court ordered Renren to pay $60,000 to Kaixinoo1, but the once-promising challenger was now a shadow of its former self. One month after that, Renren went public on the New York Stock Exchange, raising $740 million.

The lessons learned in the coliseum were clear: kill or be killed. Any company that can't fully insulate itself from competitors—on a technical, business, or even personnel level—is a target for attack. To the winner go the spoils, and those spoils can amount to billions of dollars.

It's a cultural system that also inspires a truly maniacal work ethic. Silicon Valley prides itself on long work hours, an arrangement made more tolerable by free meals, on-site gyms, and beer on tap. But compared with China's startup scene, the valley's companies look lethargic and its engineers lazy. Andrew Ng, the deep-learning pioneer who founded the Google Brain project and led AI efforts at Baidu, compared the two environments during a Sinovation event in Menlo Park:

> The pace is incredible in China. While I was leading teams in China, I'd just call a meeting on a Saturday or Sunday, or whenever I felt like it, and everyone showed up and there'd be no complaining. If I sent a text message at 7:00 PM over dinner and they haven't responded by 8:00 PM, I would wonder what's going on. It's just a constant pace of decision-making. The market does something, so you better react. That, I think, has made the China ecosystem incredible at figuring out innovations, how to take things to market. . . . I was in the US working with a vendor. I won't use any names, but a vendor I was working with actually called me up one day and they said, "Andrew, we are in Silicon Valley. You've got to stop treating us like you're in China, because we just can't deliver things at the pace you expect."

THE LEAN GLADIATOR

But the copycat era taught Chinese technology entrepreneurs more than just dirty tricks and insane schedules. The high financial stakes, propensity for imitation, and market-driven mentality also ended up incubating companies that embodied the "lean startup" methodology.

That methodology was first explicitly formulated in Silicon Valley and popularized by the 2011 book *The Lean Startup*. Core to its philosophy is the idea that founders don't know what product the market needs — the market knows what product the market needs. Instead of spending years and millions of dollars secretly creating their idea of the perfect product, startups should move quickly to release a "minimum viable product" that can tease out market demand for different functions. Internet-based startups can then receive instant feedback based on customer activity, letting them immediately begin iterating on the product: discard unused features, tack on new functions, and constantly test the waters of market demand. Lean startups must sense the subtle shifts in consumer behavior and then relentlessly tinker with products to meet that demand. They must be willing to abandon products or businesses when they don't prove profitable, pivoting and redeploying to follow the money.

By 2011, "lean" was on the lips of entrepreneurs and investors throughout Silicon Valley. Conferences and keynote speeches preached the gospel of lean entrepreneurship, but it wasn't always a natural fit for the mission-driven startups that Silicon Valley fosters. A "mission" makes for a strong narrative when pitching to media or venture-capital firms, but it can also become a real burden in a rapidly changing market. What does a founder do when there's a divergence between what the market demands and what a mission dictates?

China's market-driven entrepreneurs faced no such dilemma. Unencumbered by lofty mission statements or "core values," they had no problem following trends in user activity wherever it took their companies. Those trends often led them into industries crowded with hundreds of near-identical copycats vying for the hot market of the year. As Taobao did to eBay, these impersonators undercut any attempt to charge users by offering their own products for free. The sheer density of competition and willingness to drive prices down to zero forced companies to iterate: to tweak their products and invent new monetization models, building robust businesses with high walls that their copycat competitors couldn't scale.

In a market where copying was the norm, these entrepreneurs were forced to work harder and execute better than their opponents. Silicon Valley prides itself on its aversion to copying, but this often leads to complacency. The first mover is simply ceded a new market because others don't want to be seen as unoriginal. Chinese entrepreneurs have no such luxury. If they succeed in building a product that people want, they don't get to declare victory. They have to declare war.

WANG XING'S REVENGE

The War of a Thousand Groupons crystallized this phenomenon. Soon after its launch in 2008, Groupon became the darling of the American startup world. The premise was simple: offer coupons that worked only if a sufficient number of buyers used them. The buyers got a discount and the sellers got guaranteed bulk sales. It was a hit in post-financial-crisis America, and Groupon's valuation skyrocketed to over $1 billion in just sixteen months, the fastest pace in history.

The concept seemed tailor-made for China, where shoppers obsess over discounts and bargaining is an art form. Entrepreneurs in China looking for the next promising market quickly piled into group buying, starting local platforms based on Groupon's "Deal of the Day" model. Major internet portals launched their own group-buying divisions, and dozens of new startups entered the fray. Yet what began as dozens soon ballooned into hundreds and then thousands of copycat competitors. By the time of Groupon's initial public offering in 2011 — the largest IPO since Google's in 2004 — China was home to over five thousand different group-buying companies.

To outsiders this looked like a joke. It was a caricature of an internet ecosystem that was shameless in its copying and devoid of any original ideas. And vast swaths of those five thousand copycats were laughable, the product of ambitious but clueless entrepreneurs with no prospects for surviving the ensuing bloodletting.

But at the bottom of that dogpile, at the center of this royal rumble, was Wang Xing. In the previous seven years, he had copied three American technology products, built two companies, and sharpened the skills needed to survive in the coliseum. Wang had turned from a geeky engineer who cloned American websites into a serial entrepreneur with a keen sense for technology products, business models, and gladiatorial competition.

He put all those skills to work during the War of a Thousand Groupons. He founded Meituan ("Beautiful Group") in early 2010 and brought on battle-hardened veterans of his previous Facebook and Twitter clones to lead the charge. He didn't repeat the pixel-for-pixel copying of his Facebook and Twitter sites, instead building a user interface that better matched Chinese users' preference for densely packed interfaces.

When Meituan launched, the battle was just heating up, with competitors blowing through hundreds of millions of dollars in offline advertising. The going logic went that in order to stand out from the herd, a company had to raise lots of money and spend it to win over customers through advertising and subsidies. That high market share could then be used to raise more money and repeat the cycle. With overeager investors funding thousands of near-identical companies, Chinese urbanites took advantage of the absurd discounts to eat out in droves. It was as if China's venture-capital community were treating the entire country to dinner.

But Wang was aware of the dangers of burning cash — that's how he'd lost Xiaonei, his Facebook copy — and he foresaw the danger of trying to buy long-term customer loyalty with short-term bargains. If you only competed on subsidies, customers would endlessly jump from platform to platform in search of the best deal. Let the competitors spend the money on subsidizing meals and educating the market — he would reap the harvest that they sowed. So Wang focused on keeping costs down while iterating his product. Meituan eschewed all offline advertising, instead pouring resources into tweaking products, bringing down the cost of user acquisition and retention, and optimizing a complex back end. That back end included processing payments coming in from millions of customers and going out to tens of thousands of sellers. It was a daunting engineering challenge for which Wang's decade of hands-on experience had prepared him.

One of Meituan's core differentiations was its relationship with sellers, a crucial piece of the equation often overlooked by startups obsessed with market share. Meituan pioneered an automated payment mechanism that got money into the hands of businesses quicker, a welcome change at a time when group-buying startups were dying by the day, sticking restaurants with unpaid bills. Stability inspired loyalty, and Meituan leveraged it to build out larger networks of exclusive partnerships.

Groupon officially entered the Chinese market in early 2011 by forging a joint venture with Tencent. The marriage brought together the top international group-buying company with a homegrown giant that had both local expertise and a massive social media footprint. But the Groupon-Tencent partnership floundered from the beginning. Tencent had not yet figured out how to partner effectively with e-commerce companies, and the joint venture blindly applied Groupon's standard playbook for international expansion:

hire dozens of management consultants and use the temp agency Manpower to build out massive, low-level sales teams. Manpower headhunters made a fortune on fees, and Groupon's customer acquisition costs dwarfed those of local competitors. The foreign juggernaut was bleeding money too quickly and optimizing its product too slowly. It faded to irrelevance while the bloodletting among Chinese startups continued.

From the outside, these types of venture-funded battles for market share look to be determined solely by who can raise the most capital and thus outlast their opponents. That's half-true: while the amount of money raised is important, so is the burn rate and the "stickiness" of the customers bought through subsidies. Startups locked in these battles are almost never profitable at the time, but the company that can drive its losses-per-customer-served to the bare minimum can outlast better-funded competitors. Once the bloodshed is over and prices begin to rise, that same ruthless efficiency will be a major asset on the road to profitability.

As the War of a Thousand Groupons progressed, the combatants fought for survival in different ways. Like gladiators forming factions in the coliseum, weaker startups merged in hopes of achieving economies of scale. Others relied on bursts of high-profile advertising to briefly rise above the fray. Meituan, though, held back, consistently ranking in the top ten but not yet pushing to take the top spot.

Wang Xing embodied a philosophy of conquest tracing back to the fourteenth-century emperor Zhu Yuanzhang, the leader of a rebel army who outlasted dozens of competing warlords to found the Ming Dynasty: "Build high walls, store up grain, and bide your time before claiming the throne." For Wang Xing, venture funding was his grain, a superior product was his wall, and a billion-dollar market would be his throne.

By 2013, the dust began to settle on what had been the wildest war of copycats the country had ever seen. The vast majority of combatants had perished as victims of brutal attacks or their own mismanagement. Still standing were three gladiators: Meituan, Dianping, and Nuomi. Dianping was a longstanding Yelp copycat that had entered group buying, while Nuomi was a group-buying affiliate launched by Renren, the Facebook copycat that Wang Xing himself had founded and sold off. These three accounted for more than 80 percent of the market, and Wang's Meituan had grown to a valuation of $3 billion. After years spent photocopying American websites, he had learned the craft of the entrepreneur and won a huge chunk of a massive new market.

But it wasn't by sticking to group buying that Meituan became what it is today. Groupon had largely stayed with its original business, coasting on the novel idea of discounts through groups. By 2014, Groupon was trading at less than half of its IPO price. Today it's a shell of what it had been. By contrast, Wang ceaselessly expanded Meituan's lines of business and constantly reshaped its core products. As each hot new consumer wave washed over the Chinese economy — a booming box office, a food-delivery explosion, massive domestic tourism, flourishing online-to-offline services — Wang pivoted and ultimately transformed his company. He was voracious in his appetite for new markets and relentless in his constant iteration of new products, a prime example of a market-driven lean startup.

Meituan merged with rival Dianping in late 2015, keeping Wang in charge of the new company. By 2017 the hybrid juggernaut was fielding 20 million different orders a day from a pool of 280 million monthly active users. Most customers had long forgotten that Meituan began as a group-buying site. They knew it for what it had become: a sprawling consumer empire covering noodles, movie tickets, and hotel bookings. Today, Meituan Dianping is valued at $30 billion, making it the fourth most valuable startup in the world, ahead of Airbnb and Elon Musk's SpaceX.

ENTREPRENEURS, ELECTRICITY, AND OIL

Wang's story is about more than just the copycat who made good. His transformation charts the evolution of China's technology ecosystem, and that ecosystem's greatest asset: its tenacious entrepreneurs. Those entrepreneurs are beating Silicon Valley juggernauts at their own game and have learned how to survive in the single most competitive startup environment in the world. They then leveraged China's internet revolution and mobile internet explosion to breathe life into the country's new consumer-driven economy.

But as remarkable as these accomplishments have been, these changes will pale in comparison to what these entrepreneurs will do with the power of artificial intelligence. The dawn of the internet in China functioned like the invention of the telegraph, shrinking distances, speeding information flows, and facilitating commerce. The dawn of AI in China will be like the harnessing of electricity: a game-changer that supercharges industries across the board. The Chinese entrepreneurs who sharpened and honed their skills in the coliseum now see the power that this new technology holds, and they're already seeking out industries and applications where they can turn this energy into profit.

But to do that they need more than just their own street-smart business sensibilities. If artificial intelligence is the new electricity, big data is the oil that powers the generators. And as China's vibrant and unique internet ecosystem took off after 2012, it turned into the world's top producer of this petroleum for the age of artificial intelligence.

3

★

CHINA'S ALTERNATE
INTERNET UNIVERSE

Guo Hong is a startup founder trapped in the body of a government official. Middle-aged, Guo is always dressed in a modest dark suit and wears thick glasses. When standing for official photos at opening ceremonies, he looks no different from the dozens of other identically dressed Beijing city officials who come out to cut ribbons and deliver speeches.

During the two decades leading up to 2010, China was governed by engineers. Chinese officialdom was packed with men who studied the science of building physical things, and they put that knowledge to work transforming China from a poor agricultural society into a country of bustling factories and enormous cities. But Guo represented a new kind of official for a new era — one in which China needed to both build things and create ideas.

Put Guo alone in a room with other entrepreneurs or technologists and he suddenly comes alive. Brimming with ideas, he speaks quickly and listens intently. He has a voracious appetite for what's next in technology and an ability to envision how startups can harness these trends. Guo thinks outside the box and then takes action on the ground. He is the kind of founder that venture-capital investors love to put their money behind.

All of these habits came in handy when Guo decided to turn his slice of Beijing into the Silicon Valley of China, a hotbed for indigenous Chinese innovation. The year was 2010, and Guo was responsible for the influential Zhongguancun ("jong-gwan-soon") technology zone in northwest Beijing, an area that had long branded itself as China's answer to Silicon Valley but had not really lived up to the title. Zhongguancun was chock-full of electronics markets selling low-end smartphones and pirated software but offered few innovative startups. Guo wanted to change that.

To kick-start that process, he came to see me at the offices of my newly founded company, Sinovation Ventures. After spending a decade representing the most powerful American technology companies in China, in the fall of 2009 I left Google China to establish Sinovation, an early-stage incubator and angel investment fund for Chinese startups. I made this move because I sensed a new energy bubbling up in the Chinese startup ecosystem. The copycat era had forged world-class entrepreneurs, and they were just beginning to apply their skills to solving uniquely Chinese problems. China's rapid transition to the mobile internet and bustling urban centers created an entirely different environment, one where innovative products and new business models could thrive. I wanted to be a part of both mentoring and funding these companies as they came into their own.

When Guo came to visit Sinovation, a core team of ex-Googlers and I were working out of a small office that was located northeast of Zhongguancun. We were recruiting promising engineers to join our incubator and launch startups targeting China's first wave of smartphone users. Guo wanted to know what he could do to support that mission. I told him that the cost of rent was eating a big chunk of the money we wanted to pour into fostering these startups. Any relief on rent would mean more money for building products and companies. No problem, he said — he would make some calls. The local government could likely cover our rent for three years if we relocated to the neighborhood of Zhongguancun.

That was fantastic news for our project, and even better, Guo was just getting started. He didn't want to only throw money at one incubator. He wanted to understand what really made Silicon Valley tick. Guo began peppering me with questions about my time in the valley during the 1990s. I explained how many of the area's early entrepreneurs went on to become angel investors and mentors, how geographic proximity and tightly woven social networks gave birth to a self-sustaining venture-capital ecosystem that made smart bets on big ideas.

As we talked, I could see Guo's mind working in overdrive. He was absorbing everything and formulating the outlines of a plan. Silicon Valley's ecosystem had taken shape organically over several decades. But what if we in China could speed up that process by brute-forcing the geographic proximity? We could pick one street in Zhongguancun, clear out all the old inhabitants, and open the space to key players in this kind of ecosystem: VC firms, startups, incubators, and service providers. He already had a name in mind: Chuangye Dajie —Avenue of the Entrepreneurs.

This kind of top-down construction of an innovation ecosystem runs counter to Silicon Valley orthodoxy. In that worldview, what really makes the valley special is an abstract cultural zeitgeist, a commitment to original thinking and innovation. It's not something that could have been built merely using bricks and rent subsidies.

Guo and I both saw the value in that ethereal sense of mission, but we also saw that China was different. If we wanted to bootstrap this process in China today, money, real estate, and government support mattered. The process would require getting our hands dirty, adapting the valley's disembodied innovation ethos to the very physical realities of present-day China. The result would leverage some of the core mechanisms of Silicon Valley but would take the Chinese internet in a very different direction.

That ecosystem was becoming both independent and self-sustaining. Chinese founders no longer had to tailor their startup pitches to the tastes of foreign VCs. They could now build Chinese products to solve Chinese problems. It was a sea change that altered the very texture of the nation's cities and signaled a new era in the development of the Chinese internet. It also led to an overnight boom in production of the natural resource of the AI age.

UNCHARTED INTERNET TERRITORY

During the copycat era, the relationship between China and Silicon Valley was one of imitation, competition, and catch-up. But around 2013, the Chinese internet changed direction. It no longer lagged behind the Western internet in functionality, though it also hadn't sur-passed Silicon Valley on its own terms. Instead, it was morphing into an alternate internet universe, a space with its own raw materials, planetary systems, and laws of physics. It was a place where many users accessed the internet only through cheap smartphones, where smartphones played the role of credit cards, and where population-dense cities created a rich laboratory for blending the digital and physical worlds.

The Chinese tech companies that ruled this world had no obvi-ous corollaries in Silicon Valley. Simple shorthand like "the Amazon of China" or "the Facebook of China" no longer made sense when de-scribing apps like WeChat—the dominant social app in China, but one that evolved into a "digital Swiss Army knife" capable of letting people pay at the grocery store, order a hot meal, and book a doc-tor's visit.

Underneath this transformation lay several key building blocks: mobile-first internet users, WeChat's role as the national super-app, and mobile payments that transformed every smartphone into a digital wallet. Once those pieces were in place, Chinese startups set off an explosion of indigenous innovation. They pioneered online-to-offline services that stitched the internet deep into the fabric of the Chinese economy. They turned Chinese cities into the first cash-less environments since the days of the barter economy. And they revolutionized urban transportation with intelligent bike-sharing applications that created the world's largest internet-of-things net-work.

Adding fuel to this fire was an unprecedented wave of govern-ment support for innovation. Guo's mission to build the Avenue of the Entrepreneurs was just the first trickle of what in 2014 turned into a tidal wave of official policies pushing technology entrepre-neurship. Under the banner of "Mass Innovation and Mass Entrepre-neurship," Chinese mayors flooded their cities with new innovation zones, incubators, and government-backed venture-capital funds, many of them modeled on Guo's work with the Avenue of the Entre-preneurs. It was a campaign that analysts in the West dismissed as inefficient and misguided, but one that turbocharged the evolution of China's alternate internet universe.

Thriving in this environment required both engineering prowess and raw manpower: armies of scooter-riding deliverymen schlepping hot meals around town, tens of thousands of sales reps fanning out to push mobile payments on street vendors, and millions of shared bikes loaded onto trucks and dispersed around cities. An explosion of these services pushed Chinese companies to roll up their sleeves and do the grunt work of running an operations-heavy business in the real world.

In my view, that willingness to get one's hands dirty in the real world separates Chinese technology companies from their Silicon Valley peers. American startups like to stick to what they know: building clean digital platforms that facilitate information exchanges. Those platforms can be used by vendors who do the legwork, but the tech companies tend to stay distant and aloof from these logistical details. They aspire to the mythology satirized in the HBO series *Silicon Valley,* that of a skeleton crew of hackers building a billion-dollar business without ever leaving their San Francisco loft.

Chinese companies don't have this kind of luxury. Surrounded by competitors ready to reverse-engineer their digital products, they must use their scale, spending, and efficiency at the grunt work as a differentiating factor. They burn cash like crazy and rely on armies of low-wage delivery workers to make their business models work. It's a defining trait of China's alternate internet universe that leaves American analysts entrenched in Silicon Valley orthodoxy scratching their heads.

THE SAUDI ARABIA OF DATA

But this Chinese commitment to grunt work is also what is laying the groundwork for Chinese leadership in the age of AI implementation. By immersing themselves in the messy details of food delivery, car repairs, shared bikes, and purchases at the corner store, these companies are turning China into the Saudi Arabia of data: a country that suddenly finds itself sitting atop stockpiles of the key resource that powers this technological era. China has already vaulted far ahead of the United States as the world's largest producer of digital data, a gap that is widening by the day.

As I contended in the first chapter, the invention of deep learning means that we are moving from the age of expertise to the age of data. Training successful deep-learning algorithms requires computing power, technical talent, and lots of data. But of those three, it is the volume of data that will be the most important going forward. That's because once technical talent reaches a certain threshold, it begins to show diminishing returns. Beyond that point, data makes all the difference. Algorithms tuned by an average engineer can outperform those built by the world's leading experts if the average engineer has access to far more data.

But China's data advantage extends from quantity into quality. The country's massive number of internet users — greater than the United States and all of Europe combined — gives it the quantity of data, but it's then what those users *do* online that gives it the quality. The nature of China's alternate universe of apps means that the data collected will also be far more useful in building AI-driven companies.

Silicon Valley juggernauts are amassing data from your activity on their platforms, but that data concentrates heavily in your *online* behavior, such as searches made, photos uploaded, YouTube videos watched, and posts "liked." Chinese companies are instead gathering data from the *real* world: the what, when, and where of physical purchases, meals, makeovers, and transportation. Deep learning can only optimize what it can "see" by way of data, and China's physically grounded technology ecosystem gives these algorithms many more eyes into the content of our daily lives. As AI begins to "electrify" new industries, China's embrace of the messy details of the real world will give it an edge on Silicon Valley.

This sudden data windfall for China wasn't the result of some master plan. When Guo Hong came to see me in 2010, he couldn't have predicted the exact shape China's alternate universe would take or how machine learning would suddenly turn data into a precious commodity. But he did believe that given the right setting, funding, and a little prodding, Chinese startups could create something both totally unique and very valuable. On that point, Guo's entrepreneurial instincts were right on the money.

I left Google China and founded Sinovation Ventures a few months before Google decided to pull out of the mainland market. That move by Google was a major disappointment to our team, given the years of work we had poured into making the company competitive in China. But that departure also created an opening for Chinese startups to build an entirely new suite of products for the most exciting new trend in technology, the mobile internet.

After the iPhone's 2007 debut, the technology world began slowly adapting websites and services for access via a smartphone. In its simplest form, this meant building a version of one's website that worked well when transposed from a large computer screen onto a small smartphone. But it also meant building out new tools: an app store, photo-editing apps, and antivirus software. With Google leaving China, the market for Android-based apps in this space was now wide open. Sinovation's earliest batch of incubated startups looked to fill these gaps. In the process, I wanted us to explore a new and exciting way of interacting with the internet, a space where Silicon Valley had not yet defined the dominant paradigm.

During China's copycat era, the small portion of its population that accessed the internet did so in the same way as Americans, through a desktop or laptop computer. Chinese users' behavior differed significantly from that of Americans, but the fundamental tools used were the same. Computers were still too expensive for most Chinese people, and by 2010 only around one-third of China's population had access to the internet. So when cheap smartphones hit the market, waves of ordinary citizens leapfrogged over personal computers entirely and went online for the first time via their phones.

Simple as that transition sounds, it had profound implications for the particular shape that the Chinese internet would take. Smartphone users not only acted differently than their desktop peers; they also wanted different things. For mobile-first users, the internet wasn't just an abstract collection of digital information that you accessed from a set location. Rather, the internet was a tool that you brought with you as you moved around cities — it should help solve the local problems you run into when you need to eat, shop, travel, or just get across town. Chinese startups needed to build their products accordingly.

This opened a real opportunity for Chinese startups backed by Chinese VCs to break new ground in order to foster Chinese-style innovation. At Sinovation, our first round of investment went into incubating nine companies, several of which were eventually acquired or controlled by Baidu, Alibaba, and Tencent. Those three Chinese internet juggernauts (collectively known by the abbreviation "BAT") used our startups to accelerate their transition into mobile internet companies. Those startup acquisitions formed a solid foundation for their mobile efforts, but it would be a secretive in-house project at Tencent that first cracked open the potential of what I call China's alternate internet universe.

WECHAT: HUMBLE BEGINNINGS, HUGE AMBITIONS

Hardly anyone noticed when the world's most powerful app waltzed onto the world stage. The January 2011 launch of WeChat, Tencent's new social messaging app, received only one mention in the English-language press, on the technology site the Next Web. Tencent already owned the two dominant social networks in China — its QQ instant messaging platform and Q-Zone social network each had hundreds of millions of users — but American analysts dismissed these as mediocre knockoffs of American products. The company's new smartphone app didn't even have an English name yet, going only by the Chinese name Weixin, or "micro-message."

But it did have a few other things going for it. The app lets you send photos and short voice recordings along with typing out messages. The latter was a major benefit given how cumbersome inputting Chinese characters on a phone was at the time. WeChat was also created specifically for smartphones. Instead of trying to transform its dominant desktop platform, QQ, into a phone app, Tencent aimed to disrupt its own product with a better one built just for mobile. It was a risky strategy for an established juggernaut, but one that paid off big time.

The app's clean functionality took off, and as WeChat gained users, it also tacked on more functions. In just over a year it had hit 100 million registered users, and by its two-year anniversary in January 2013, that number was 300 million. Along the way it had added voice and video calls and conference calls, functions that seem obvious today but that WeChat's global competitor WhatsApp waited until 2016 to incorporate.

WeChat's early tweaks and optimizations were just the beginning. It soon pioneered an innovative "app-within-an-app" model that changed the way media outlets and advertisers used social platforms. These were WeChat's "official accounts," subscription-based third-party content streams that lived within the app and were sometimes compared to Facebook pages for media companies. But instead of Facebook's minimalist platform for posting content, the official accounts offered much of the functionality of a standalone app without the hassle of actually building one. These accounts quickly became so dominant in the social media space that many media and consumer companies simply stopped building their own apps, choosing instead to live entirely in WeChat's world.

In the span of two years, WeChat went from a no-name app to a powerhouse of messaging, media, marketing, and gaming. But Tencent wanted even more. It already monopolized users' digital lives, but it wanted to extend that functionality beyond the smartphone.

Over the ensuing five years, Tencent painstakingly built WeChat into the world's first super-app. It became a "remote control for life" that dominated not just users' digital worlds but allowed them to pay at restaurants, hail taxis, unlock shared bikes, manage investments, book doctors' appointments, and have those doctors' prescriptions delivered to your door. This metastasizing functionality would blur the lines dividing our online and offline worlds, both molding and feeding off of China's alternate internet universe. But before it could do that, WeChat had to get inside its users' wallets, and that meant taking on the top dog in digital commerce.

The attack came on the most festive night of the Chinese calendar — Chinese New Year's Eve, 2014 — and the weapon drew inspiration from the occasion. Chinese tradition calls for the gifting of "red envelopes" during Chinese New Year, small and decorative red packets with cash inside. That cash is the Chinese equivalent of a Christmas present, something usually given by older relatives to children, and by bosses to employees.

Tencent's innovation was so simple — and such pure fun for users — that it masked the magnitude of the power grab. WeChat gave its users the ability to send out digital red envelopes containing real money to WeChat friends near and far. Once users linked their bank accounts to WeChat, they could send out envelopes worth a set amount of money to one person or into a group chat and let their friends race to see who could "open" it first and get the money. That money then lived inside users' WeChat Wallet, a new subdivision of the app. The money could be used to make purchases, transferred to other friends, or added to their own bank account if they linked it with WeChat.

It was a seamless translation to digital of an age-old Chinese tradition, one that added a gaming element to the process. WeChat users loved the envelopes, sending out 16 million of the packets during Chinese New Year and in the process, linking 5 million new bank accounts to WeChat Wallet.

Jack Ma was less amused. He called the move by Tencent a "Pearl Harbor attack" on Alibaba's dominance in digital commerce. Alibaba's Alipay had pioneered digital payments tailored for Chinese users back in 2004 and later adapted the product for smartphones. But overnight WeChat had taken all the momentum in new types of mobile payments, nudging millions of new users into linking their bank accounts to what was already the most powerful social app in China. Ma warned Alibaba employees that if they didn't fight to hold their grip on mobile payments, it would spell the company's end. Observers at the time thought this was just typical over-the-top rhetoric from Jack Ma, a charismatic entrepreneur with a genius for rallying his troops. But looking back four years later, it seems likely that Ma saw what was coming.

The four years leading up to Tencent's Pearl Harbor moment saw many of the pieces of China's alternate internet universe fall into place. Gladiatorial competition between China's copycat start-ups had trained a generation of street-smart internet entrepreneurs. Smartphone users had more than doubled between 2009 and 2013, from 233 million to a whopping 500 million. Early-stage funds were fostering a new generation of startups building innovative mobile apps for this market. And WeChat demonstrated the power of the super-app installed on virtually everyone's smartphone, an all-in-one portal to the Chinese mobile ecosystem.

When Tencent's flood of red envelopes lured millions of Chinese into linking their bank accounts to WeChat, it put in place the last crucial puzzle piece of a consumption revolution: the ability to pay for anything and everything with your phone. Over the coming years, Alibaba, Tencent, and thousands of Chinese startups would race to apply these tools to every nook and cranny of Chinese urban life, including food delivery, electricity bills, live-streaming celebrities, on-demand manicures, shared bikes, train tickets, movie tickets, and traffic tickets. China's online and offline world would begin rubbing shoulders in a way not seen anywhere else in the world. They were refashioning China's urban landscape and the world's richest real-world datascape.

But building an alternate internet universe that reaches into every corner of the Chinese economy couldn't be done without the country's most important economic actor: the Chinese government.

IF YOU BUILD IT, THEY WILL COME

On that front, Guo Hong was ahead of the curve. In the years after his first visit to my office, his dream of an Avenue of the Entrepreneurs had been turned into a plan, and that plan turned into action. Guo chose for his experiment a pedestrian street in Zhongguancun that was home to a mishmash of bookstores, restaurants, and knockoff electronics markets.

Back in the 1980s, the government had already transformed this street for the sake of an economic upgrade. At the time, China was in the throes of export-driven growth and urbanization, two projects that required engineering expertise that the country lacked. So officials turned the walking street into a "Book City" packed with stores carrying modern science and engineering textbooks for students at nearby Tsinghua and Peking University to pore over. By the year 2010, the rise of the Chinese internet had driven many of the bookstores out of business, replacing them with small storefronts hawking cheap electronics and pirated software — the raw ingredients of China's copycat era.

But Guo wanted to turbocharge an upgrade to a new era of indigenous innovation. His original small-scale experiment in attracting Sinovation Ventures via rent subsidies had succeeded, and so Guo planned to refurbish an entire street for high-tech tenants. He and the local district government used a combination of cash subsidies and offers of space elsewhere to move out almost all the traditional businesses on the street. In 2013, construction crews took jackhammers and paving equipment to the now-empty street, and after a year of laying bricks and building sleek new exteriors, on June 11, 2014, the Avenue of the Entrepreneurs opened to its new tenants.

Guo had used the tools at his disposal — cash, cement, and manual labor — to give a strong nudge toward indigenous innovation in the local startup. It was a landmark moment for Zhongguancun, but one that wasn't destined to stay sequestered to this corner of Beijing. Indeed, Guo's approach was about to go national.

INNOVATION FOR THE MASSES

On September 10, 2014, Premier Li Keqiang took the stage during the 2014 World Economic Forum's "Summer Davos" in the coastal Chinese city of Tianjin. There he spoke of the crucial role technological innovation played in generating growth and modernizing the Chinese economy. The speech was long and dense, heavy on jargon and light on specifics. But of note during the speech, Li repeated a phrase that was new to the Chinese political lexicon: "mass entrepreneurship and mass innovation." He concluded by wishing the attendees a successful forum and good health.

To outside observers, it was an utterly unremarkable event, and there was almost no coverage in the Western press. Chinese leaders deliver speeches like this almost every day, long, plodding, and full of stock phrases that ring hollow to Western ears. Those phrases can act as signals during internal debates within the Chinese government, but they don't necessarily translate to immediate changes in the real world.

This time was different. Li's speech lit the first spark of what would become a raging fire in the Chinese technology industry, pushing activity in the investment and startup space to feverish new heights. The new phrase — "mass entrepreneurship and mass innovation" — became the slogan for a momentous government push to foster startup ecosystems and support technological innovation. Guo Hong's proactive approach to innovation was suddenly being scaled up across the world's second-largest economy, and it would turbocharge the creation of the only true counterweight to Silicon Valley.

China's mass innovation campaign did that by directly subsidizing Chinese technology entrepreneurs and shifting the cultural zeitgeist. It gave innovators the money and space they needed to work their magic, and it got their parents to finally stop nagging them about taking a job at a local state-owned bank.

Nine months after Li's speech, China's State Council — roughly equivalent to the U.S. president's cabinet — issued a major directive on advancing mass entrepreneurship and innovation. It called for the creation of thousands of technology incubators, entrepreneurship zones, and government-backed "guiding funds" to attract greater private venture capital. The State Council's plan promoted preferential tax policies and the streamlining of government permits for starting a business.

China's central government laid out the goals, but implementation was left up to thousands of mayors and local officials scattered around the country. Promotion for local officials in China's government bureaucracy is based on performance evaluations conducted by higher-ups within the Communist Party's internal human resources department. So when the central government sets a clear goal — a new metric on which lower-level officials can demonstrate

their competence — ambitious officials everywhere throw themselves into advancing that goal and proving themselves capable.

Following the issuance of the State Council directive, cities
around China rapidly copied Guo Hong's vision and rolled out their
own versions of the Avenue of the Entrepreneurs. They used tax discounts and rent rebates to attract startups. They created one-stop-
shop government offices where entrepreneurs could quickly register
their companies. The flood of subsidies created 6,600 new startup
incubators around the nation, more than quadrupling the overall
total. Suddenly, it was easier than ever for startups to get quality
space, and they could do so at discount rates that left more money
for building their businesses.

Larger city and provincial governments pioneered different models for "guiding funds," a mechanism that uses government money
to spur more venture investing. The funds do that by increasing the
upside for private investors without removing the risk. The government uses money from the guiding fund to invest in private venture-
capital funds in the same role as other private limited partners. If the
startups that fund invested in (the "portfolio companies") fail, all the
partners lose their investment, including the government.

But if the portfolio companies succeed — say, double in value
within five years — then the fund's manager caps the government's
upside from the fund at a predetermined percentage, perhaps 10
percent, and uses private money to buy the government's shares
out at that rate. That leaves the remaining 90 percent gain on the
government's investment to be distributed among private investors
who have already seen their own investments double. Private investors are thus incentivized to follow the government's lead, investing
in funds and industries that the local government wants to foster.
During China's mass innovation push, use of local government guiding funds exploded, nearly quadrupling from $7 billion in 2013 to $27
billion in 2015.

Private venture funding followed. When Sinovation was founded
in 2009, China was experiencing such rapid growth in manufacturing and real estate that the smart money was still pouring into those
traditional sectors. But in 2014, this all turned around. For three of

the four years leading up to 2014, total Chinese VC funding held steady at around $3 billion. In 2014, that immediately quadrupled to $12 billion, and then doubled again to $26 billion in 2015. Now it seemed like any smart and experienced young person with a novel idea and some technical chops could throw together a business plan and find funding to get his or her startup off the ground.

American policy analysts and investors looked askance at this heavy-handed government intervention in what are supposed to be free and efficient markets. Private-sector players make better bets when it comes to investing, they said, and government-funded innovation zones or incubators will be inefficient, a waste of taxpayer money. In the minds of many Silicon Valley power players, the best thing that the federal government can do is leave them alone.

But what these critics miss is that this process can be both highly inefficient and extraordinarily effective. When the long-term upside is so monumental, overpaying in the short term can be the right thing to do. The Chinese government wanted to engineer a fundamental shift in the Chinese economy, from manufacturing-led growth to innovation-led growth, and it wanted to do that in a hurry.

It could have taken a hands-off approach, standing aside while investment returns in traditional industries fell and private investment slowly made its way into the high-tech sector. That shift would be subject to the ordinary frictions of human endeavors: imperfect information, old-school investors who weren't so sure about this internet thing, and plain old economic inertia. Eventually, though, those frictions would be overcome, and money would make its way into private venture funds that might spend each dollar more efficiently than the government could.

But that's a process that would take many years, if not decades. China's top leadership did not have the patience to wait. It wanted to use government money to brute-force a faster transformation, one that would pay dividends through an earlier transition to higher-quality growth. That process of pure force was often locally inefficient — incubators that went unoccupied and innovation avenues that never paid off — but on a national scale, the impact was tremendous.

The effects of China's mass entrepreneurship and mass innovation campaign went far beyond mere office space and investment dollars. The campaign left a deep imprint on ordinary people's perceptions of internet entrepreneurship, genuinely shifting the cultural zeitgeist.

Chinese culture traditionally has a tendency toward conformity and a deference toward authority figures, such as parents, bosses, teachers, and government officials. Before a new industry or activity has received the stamp of approval from authority figures, it's viewed as inherently risky. But if that industry or activity receives a ringing endorsement from Chinese leadership, people will rush to get a piece of the action. That top-down structure inhibits free-ranging or exploratory innovation, but when the endorsement arrives and the direction is set, all corners of society simultaneously spring into action.

Before 2014, the Chinese government had never made clear exactly how it viewed the rise of the Chinese internet. Despite the early successes of companies like Baidu and Alibaba, periods of relative openness online were followed by ominous signals and legal crackdowns on users "spreading rumors" via social media platforms. No one could be sure what was coming next. With the mass innovation campaign, the Chinese government issued its first full-throated endorsement of internet entrepreneurship. Posters and banners sprung up around the country exhorting everyone to join the cause. Official media outlets ran countless stories touting the virtues of indigenous innovation and trumpeting the successes of homegrown startups. Universities raced to offer new courses around entrepreneurship, and bookstores filled up with biographies of tech luminaries and self-help books for startup founders.

Throwing even more fuel on this fire was Alibaba's record-breaking 2014 debut on the New York Stock Exchange. A group of Taobao sellers rang the opening bell for Alibaba's initial public offering on September 19, just nine days after Premier Li's speech. When the dust settled on a furious round of trading, Alibaba had claimed the title of the largest IPO in history, and Jack Ma was crowned the richest man in China.

But it was about more than just the money. Ma had become a national hero, but a very relatable one. Blessed with a goofy charisma, he seems like the boy next door. He didn't attend an elite university and never learned how to code. He loves to tell crowds that when KFC set up shop in his hometown, he was the only one out of twenty-five applicants to be rejected for a job there. China's other early internet giants often held Ph.D.s or had Silicon Valley experience in the United States. But Ma's ascent to rock-star status gave a new meaning to "mass entrepreneurship" — in other words, this was something that anyone from the Chinese masses had a shot at.

The government endorsement and Ma's example of internet entrepreneurship were particularly effective at winning over some of the toughest customers: Chinese mothers. In the traditional Chinese mentality, entrepreneurship was still something for people who couldn't land a real job. The "iron rice bowl" of lifetime employment in a government job remained the ultimate ambition for older generations who had lived through famines. In fact, when I had started Sinovation Ventures in 2009, many young people wanted to join the startups we funded but felt they couldn't do so because of the steadfast opposition of their parents or spouses. To win these families over, I tried everything I could think of, including taking the parents out to nice dinners, writing them long letters by hand, and even running financial projections of how a startup could pay off. Eventually we were able to build strong teams at Sinovation, but every new recruit in those days was an uphill battle.

By 2015, these people were beating down our door — in one case, literally breaking Sinovation's front door — for the chance to work with us. That group included scrappy high school dropouts, brilliant graduates of top universities, former Facebook engineers, and more than a few people in questionable mental states. While I was out of town, the Sinovation headquarters received a visit from one would-be entrepreneur who refused to leave until I met with him. When the staff told him that I wouldn't be returning any time soon, the man lay on the ground and stripped naked, pledging to lie right there until Kai-Fu Lee listened to his idea.

That particular entrepreneur received a police escort rather than a seed investment, but the episode captures the innovation mania that was gripping China. A country that had spent a decade dancing around the edges of internet entrepreneurship was now plunging in headfirst. The same went for Guo Hong. While creating the Avenue of the Entrepreneurs, Guo caught the entrepreneurial bug himself, and in 2017 he left the world of Chinese officialdom to become the founder and chairman of Zhongguancun Bank, a financial "startup" modeled on Silicon Valley Bank and dedicated to serving local entrepreneurs and innovators.

All the pieces were now in place for the flourishing of China's alternate internet universe. It had the leapfrog technology, the funding, the facilities, the talent, and the environment. The table was set to create internet companies that were new, valuable, and uniquely Chinese.

HERE, THERE, AND O2O EVERYWHERE

To do all of this, the Chinese internet had to get its hands dirty. For two decades, Chinese internet companies had played a role similar to that of their American peers: information nodes on a digital network. Now they were ready to dive into the nitty-gritty details of daily life.

Analysts dubbed the explosion of real-world internet services that blossomed across Chinese cities the "O2O Revolution," short for "online-to-offline." The terminology can be confusing but the concept is simple: turn online actions into offline services. E-commerce websites like Alibaba and Amazon had long done this for the purchase of durable physical goods. The O2O revolution was about bringing that same e-commerce convenience to the purchase of real-world services, things that can't be put in a cardboard box and shipped across country, like hot food, a ride to the bar, or a new haircut.

Silicon Valley gave birth to one of the first transformational O2O models: ride-sharing. Uber used cell phones and personal cars to change how people got around cities in the United States and then around the world. Chinese companies like Didi Chuxing quickly copied the business model and adapted it to local conditions, with Didi eventually driving Uber out of China and now battling it in global markets. Uber may have given an early glimpse of O2O, but it was Chinese companies that would take the core strengths of that model and apply it to transforming dozens of other industries.

Chinese cities were the perfect laboratory for experimentation. Urban China can be a joy, but it can also be a jungle: crowded, polluted, loud, and less than clean. After a day spent commuting on crammed subways and navigating eight-lane intersections, many middle-class Chinese just want to be spared another trip outdoors to get a meal or run an errand. Lucky for them, these cities are also home to large pools of migrant laborers who would gladly bring that service to their door for a small fee. It's an environment built for O2O.

The first O2O service other than ride-hailing to truly take off was food delivery. China's internet juggernauts and a flood of start-ups like Wang Xing's Meituan Dianping all made O2O food delivery plays, pouring subsidies and engineering resources into the market. Crowds at Chinese restaurants thinned out, and streets filled up with swarms of electric scooters trailing steam from the hot meals they carried on board. Payments could be made seamlessly through We-Chat Wallet and Alipay. By the end of 2014, Chinese spending on O2O food delivery had grown by over 50 percent and topped 15 billion RMB. By 2016, China's 20 million daily online food orders equaled ten times the total across the United States.

From there the O2O models became even more creative. Some hair stylists and manicurists gave up their storefronts entirely, exclusively booking through apps and making house calls. People who were feeling ill could hire others to wait in the famously long lines outside hospitals. Lazy pet owners could use an app to hail someone who would come right over and clean out a cat's litter box or wash their dog. Chinese parents could hire van drivers to pick up their children from school, confirming their ID and arrival home through apps. Those who didn't want to have children could use another app for around-the-clock condom delivery.

For Chinese people, the transition took the edge off urban life. For small businesses, it meant a boom in customers, as the reductions in friction led Chinese urbanites to spend more. And for China's new wave of startups, it meant skyrocketing valuations and a ceaseless drive to push into ever more sectors of urban life.

After a couple of years of explosive growth and gladiatorial competition, the manic production of new O2O models cooled off. Many overnight O2O unicorns died once the subsidy-fueled growth ended. But the innovators and gladiators who survived — like Wang Xing's Meituan Dianping — multiplied their already billion-dollar valuations by fundamentally reshaping urban China's service sector. By late 2017, Meituan Dianping was valued at $30 billion, and Didi Chuxing hit a valuation of $57.6 billion, surpassing that of Uber itself.

It was a social and commercial transformation that was powered by — and which further empowered — WeChat. Installed on more than half of all smartphones in China and now linked to many users' bank accounts, WeChat had the power to nudge hundreds of millions of Chinese into O2O purchases and to pick winners among the competing startups. WeChat Wallet linked up with top O2O startups so that WeChat users could hail a taxi, order a meal, book a hotel, manage a phone bill, and buy a flight to the United States, all without ever leaving the app. (Not coincidentally, most of the startups WeChat picked to feature in its Wallet were also the recipients of Tencent investments.)

With the rise of O2O, WeChat had grown into the title bestowed on it by Connie Chan of leading VC fund Andreesen Horowitz: a remote control for our lives. It had become a super-app, a hub for diverse functions that are spread across dozens of different apps in other ecosystems. In effect, WeChat has taken on the functionality of Facebook, iMessage, Uber, Expedia, eVite, Instagram, Skype, PayPal, Grubhub, Amazon, LimeBike, WebMD, and many more. It isn't a perfect substitute for any one of those apps, but it can perform most of the core functions of each, with frictionless mobile payments already built in.

This all marks a stark contrast to the "app constellation" model in Silicon Valley in which each app sticks to a strictly prescribed set of functions. Facebook even went so far as to split its social network and messaging functions into two different apps, Facebook and Messenger. Tencent's choice to go for the super-app model appeared risky at the start: could you possibly bundle so many things together without overwhelming the user? But the super-app model proved wildly successful for WeChat and has played a crucial role in shaping this alternate universe of internet services.

THE LIGHT TOUCH VERSUS HEAVYWEIGHTS

But the O2O revolution showcased an even deeper — and in the age of AI implementation, more impactful — divide between Silicon Valley and China — what I call "going light" versus "going heavy." The terms refer to how involved an internet company becomes in providing goods or services. They represent the extent of vertical integration as a company links up the on- and offline worlds.

When looking to disrupt a new industry, American internet companies tend to take a "light" approach. They generally believe the internet's fundamental power is sharing information, closing knowledge gaps, and connecting people digitally. As internet-driven companies, they try to stick to this core strength. Silicon Valley startups will build the information platform but then let brick-and-mortar businesses handle the on-the-ground logistics. They want to win by outsmarting opponents, by coming up with novel and elegant code-based solutions to information problems.

In China, companies tend to go "heavy." They don't want to just build the platform — they want to recruit each seller, handle the goods, run the delivery team, supply the scooters, repair those scooters, and control the payment. And if need be, they'll subsidize that entire process to speed user adoption and undercut rivals. To Chinese startups, the deeper they get into the nitty-gritty — and often very expensive — details, the harder it will be for a copycat competitor to mimic the business model and undercut them on price. Going heavy means building walls around your business, insulating yourself from the economic bloodshed of China's gladiator wars. These companies win both by outsmarting their opponents and by outworking, outhustling, and outspending them on the street.

It's a distinction captured well by comparing well-known restaurant platforms in two countries, Yelp and Dianping. Both were founded around 2004 as desktop platforms for posting restaurant reviews. They both eventually became smartphone apps, but while Yelp largely stuck to reviews, Dianping dove headfirst into the group-buying frenzy: building out payments, developing vendor relationships, and spending massively on subsidies.

When the two companies went into online ordering and delivery, they took different approaches. Yelp moved late and went light. After eleven years as a purely digital platform that lived off advertising, in 2015 Yelp finally took a baby step into deliveries by acquiring Eat24, an ordering and food-delivery platform. But it still asked restaurants to handle the majority of deliveries, just using Eat24 to fill in gaps for restaurants that didn't have delivery teams. The lightweight process offered restaurants few real incentives to participate, and as a result, the business never fully took off. Within two and a half years, Yelp had given up, selling Eat24 to Grubhub and retreating to its lightweight approach. "[The sale to Grubhub] allowed us to do what we do best," explained Yelp CEO Jeremy Stoppelman, "which was to build the Yelp app."

In contrast, Dianping went into commerce early and went very heavily into food delivery. After four years in the trenches of the group-buying wars, Dianping began piloting food delivery in late 2013. It spent millions of dollars hiring and managing fleets of scooter-riding teams that delivered orders from restaurant to doorstep. Dianping's delivery teams did the legwork, so every mom-and-pop shop suddenly had the option of expanding its customer base without having to hire a delivery team.

By throwing tons of money and people at the problem, Dianping could attain economies of scale in China's dense urban centers. It was an expensive and logistically taxing endeavor, but one that ultimately improved efficiency and reduced costs for the end customer. Eighteen months after debuting its delivery service, Dianping doubled down on those economies of scale by merging with archrival Meituan. By 2017, Meituan Dianping's valuation of $30 billion was more than *triple* that of Yelp and Grubhub combined.

Other examples of O2O companies in China going heavy abound. After driving Uber out of the Chinese ride-hailing market, Didi has begun buying up gas stations and auto repair shops to service its fleet, making great margins because of its understanding of its drivers and their trust in the Didi brand. While Airbnb largely remains a lightweight platform for listing your home, the company's Chinese rival, Tujia, manages a large chunk of rental properties itself. For Chinese hosts, Tujia offers to take care of much of the grunt work: cleaning the apartment after each visit, stocking it with supplies, and installing smart locks.

That willingness to go heavy — to spend the money, manage the workforce, do the legwork, and build economies of scale — has reshaped the relationship between the digital and real-world economies. China's internet is penetrating far deeper into the economic lives of ordinary people, and it is affecting both consumption trends and labor markets. In a 2016 study by McKinsey and Company, 65 percent of Chinese O2O users said that the apps led them to spend more money on dining. In the categories of travel and transportation, 77 percent and 42 percent of users, respectively, reported increasing their spending.

In the short run, this cash-flow stimulated the Chinese economy and pumped up valuations. But the long-term legacy of this movement is the data environment it created. By enrolling the vendors, processing the orders, delivering the food, and taking in the payments, China's O2O champions began amassing a wealth of real-world data on the consumption patterns and personal habits of their users. Going heavy gave these companies a data edge over their Silicon Valley peers, but it was mobile payments that would extend their reach even further into the real world and turn that data edge into a commanding lead.

SCAN OR GET SCANNED

As O2O spending exploded, Alipay and Tencent decided to make a direct bid for disrupting the country's all-cash economy. (In 2011, Alibaba spun off its financial services, including Alipay, into a company that would become Ant Financial.) China had never fully embraced credit and debit cards, instead sticking to cash for the vast majority of all transactions. Large supermarkets or shopping malls let customers swipe a card, but the mom-and-pop shops and family restaurants that dominate the cityscape rarely had point-of-sale (POS) devices for processing plastic cards.

The owners of those shops did, however, have smartphones. So China's internet juggernauts turned those phones into mobile portals for payments. The idea was simple, but the speed of execution, impact on consumer behavior, and resulting data have been astonishing.

During 2015 and 2016, Tencent and Alipay gradually introduced the ability to pay at shops by simply scanning a QR code — basically a square bar code for phones — within the app. It's a scan-or-get-scanned world. Larger businesses bought simple POS devices that can scan the QR code displayed on customers' phones and charge them for the purchase. Owners of small shops could just print out a picture of a QR code that was linked to their WeChat Wallet. Customers then use the Alipay or WeChat apps to scan the code and enter the payment total, using a thumbprint for confirmation. Funds are instantly transferred from one bank account to the other — no fees and no need to fumble with wallets. It marked a stark departure from the credit-card model in the developed world. When they were first introduced, credit cards were cutting edge, the most convenient and cost-effective solution to the payment problem. But that advantage has now turned into a liability, with fees of 2.5 to 3 percent on most charges turning into a drag on adoption and utilization.

China's mobile payment infrastructure extended its usage far beyond traditional debit cards. Alipay and WeChat even allow peer-to-peer transfers, meaning you can send money to family, friends, small-time merchants, or strangers. Frictionless and hooked into mobile, the apps soon turned into tools for "tipping" the creators of online articles and videos. Micro-payments of as little as fifteen cents flourished. The companies also decided not to charge commissions on the vast majority of transfers, meaning people accepted mobile payments for all transactions — none of the mandatory minimum purchases or fifty-cent fees charged by U.S. retailers on small purchases with credit cards.

Adoption of mobile payments happened at lightning speed. The two companies began experimenting with payment-by-scan in 2014 and deployed at scale in 2015. By the end of 2016, it was hard to find a shop in a major city that did not accept mobile payments. Chinese people were paying for groceries, massages, movie tickets, beer, and bike repairs within just these two apps. By the end of 2017, 65 percent of China's over 753 million smartphone users had enabled mobile payments.

Given the extremely low barriers to entry, those payment systems soon trickled down into China's vast informal economy. Migrant workers selling street food simply let customers scan and send over payments while the owner fried the noodles. It got to the point where beggars on the streets of Chinese cities began hanging pieces of paper around their necks with printouts of two QR codes, one for Alipay and one for WeChat.

Cash has disappeared so quickly from Chinese cities that it even "disrupted" crime. In March 2017, a pair of Chinese cousins made headlines with a hapless string of robberies. The pair had traveled to Hangzhou, a wealthy city and home to Alibaba, with the goal of making a couple of lucrative scores and then skipping town. Armed with two knives, the cousins robbed three consecutive convenience stores only to find that the owners had almost no cash to hand over — virtually all their customers were now paying directly with their phones. Their crime spree netted them around $125 each — not even enough to cover their travel to and from Hangzhou — when police picked them up. Local media reported rumors that upon arrest one of the brothers cried out, "How is there no cash left in Hangzhou?"

It made for a sharp contrast with the stunted growth of mobile payments in the United States. Google and Apple have taken a stab at mobile payments with Google Wallet and Apple Pay, but neither has really attained widespread adoption. Apple and Google don't release user figures for their platforms, but everyday observation and more rigorous analysis both point to massive gaps in adoption. The market research firm iResearch estimated in 2017 that Chinese mobile payment spending outnumbered that in the United States by a ratio of fifty to one. For 2017, total transactions on China's mobile payment platforms reportedly surpassed $17 trillion — greater than China's GDP — an astounding number made possible by the fact that these payments allow for peer-to-peer transfers and multiple mobile transactions for items and services throughout the chain of production.

LEAPING FROGS AND TAXI DRIVERS

That massive gap is partly explained by the strength of the incumbent. Americans already benefit from (and pay for) the convenience of credit and debit cards — the cutting-edge financial technology of the 1960s. Mobile payments are an improvement on cards but not as dramatic an improvement as the jump straight from cash. As with China's rapid transition to the mobile internet, the country's weakness in incumbent technology (desktop computers, landline phones, and credit cards) turned into the strength that let it leapfrog into a new paradigm.

But that leap to mobile payments wasn't just a product of weak incumbents and independent consumer choices. Alibaba and Tencent accelerated the transition by forcing adoption through massive subsidies, a form of "going heavy" that makes American technology companies squirm.

In the early days of ride-hailing apps in China, riders could book through apps but often paid in cash. A large portion of cars on the leading Chinese platforms were traditional taxis driven by older men — people who weren't in a rush to give up good old cash. So Tencent offered subsidies to both the rider and the driver if they used WeChat Wallet to pay. The rider paid less and the driver received more, with Tencent making up the difference for both sides.

The promotion was extremely costly — due to both legitimate rides and fraudulent ones designed to milk subsidies — but Tencent persisted. That decision paid off. The promotion built up user habits and lured onto the platform taxi drivers, who are the key nodes in the urban consumer economy.

By contrast, Apple Pay and Google Wallet have tread lightly in this arena. They theoretically offer greater convenience to users, but they haven't been willing to bribe users into discovering that method for themselves. Reluctance on the part of U.S. tech giants is understandable: subsidies eat into quarterly revenue, and attempts to "buy users" are usually frowned on by Silicon Valley's innovation purists.

But that American reluctance to go heavy has slowed adoption of mobile payments and will hurt these companies even more in a data-driven AI world. Data from mobile payments is currently generating the richest maps of consumer activity the world has ever known, far exceeding the data from traditional credit-card purchases or online activity captured by e-commerce players like Amazon or platforms like Google and Yelp. That mobile payment data will prove invaluable in building AI-driven companies in retail, real estate, and a range of other sectors.

BEIJING BICYCLE REDUX

While mobile payments totally transformed China's financial landscape, shared bicycles transformed its urban landscapes. In many ways, the shared bike revolution was turning back the clock. From the time of the Communist Revolution in 1949 through the turn of the millennium, Chinese cities were teeming with bicycles. But as economic reforms created a new middle class, car ownership took off and riding a bicycle became something for individuals who were too poor for four-wheeled transport. Bikes were pushed to the margins of city streets and the cultural mainstream. One woman on the country's most popular dating show captured the materialism of the moment when she rejected a poor suitor by saying, "I'd rather cry in the back of a BMW than smile on the back of a bicycle."

And then, suddenly, China's alternate universe reversed the tide. Beginning in late 2015, bike-sharing startups Mobike and ofo started supplying tens of millions of internet-connected bicycles and distributing them around major Chinese cities. Mobike outfitted its bikes with QR codes and internet-connected smart locks around the bike's back wheel. When riders use the Mobike app (or its mini-app in We-Chat Wallet) to scan a bike's QR code, the lock on the back wheel automatically slides open. Mobike users ride the bike anywhere they want and leave it there for the next rider to find. Costs of a ride are based on distance and time, but heavy subsidies mean they often come in at 15 cents or less. It's a revolutionary, real-world innovation, one made possible by mobile payments. Adding credit-card POS machines to bikes would be too expensive and repair-intensive, but frictionless mobile payments are both cheap to layer onto a bike and incredibly efficient.

Shared-bike use exploded. In the span of a year, the bikes went from urban oddities to total ubiquity, parked at every intersection, sitting outside every subway exit, and clustered around popular shops and restaurants. It rarely took more than a glance in either direction to find one, and five seconds in the app to unlock it. City streets turned into a rainbow of brightly colored bicycles: orange and silver for Mobike; bright yellow for ofo; and a smattering of blue, green, and red for other copycat companies. By the fall of 2017, Mobike was logging 22 million rides per day, almost all of them in China. That is four times the number of *global* rides Uber was giving each day in 2016, the last time it announced its totals. In the spring of 2018, Mobike was acquired by Wang Xing's Meituan Dianping for $2.7 billion, just three years after the bike-sharing company's founding.

Something new was emerging from all those rides: perhaps the world's largest and most useful internet-of-things (IoT) networks. The IoT refers to collections of real-world, internet-connected devices that can convey data from the world around them to other devices in the network. Most Mobikes are equipped with solar-powered GPS, accelerators, Bluetooth, and near-field communications capabilities that can be activated by a smartphone. Together, those sensors generate twenty terabytes of data per day and feed it all back into Mobike's cloud servers.

BLURRED LINES AND BRAVE NEW WORLDS

In the span of less than two years, China's bike-sharing revolution has reshaped the country's urban landscape and deeply enriched its data-scape. This shift forms a dramatic visual illustration of what China's alternate internet universe does best: solving practical problems by blurring the lines between the online and offline worlds. It takes the core strength of the internet (information transmission) and leverages it in building businesses that reach out into the real world and directly touch on every corner of our lives.

Building this alternate universe didn't happen overnight. It required market-driven entrepreneurs, mobile-first users, innovative super-apps, dense cities, cheap labor, mobile payments, and a government-sponsored culture shift. It's been a messy, expensive, and disruptive process, but the payoff has been tremendous. China has built a roster of technology giants worth over a trillion dollars — a feat accomplished by no other country outside the United States.

But the greatest riches of this new Chinese tech world have yet to be realized. Like the long-buried organic matter that became fossil fuels powering the Industrial Revolution, the rich real-world interactions in China's alternate internet universe are creating the massive data that will power its AI revolution. Each dimension of that universe — WeChat activity, O2O services, ride-hailing, mobile payments, and bike-sharing — adds a new layer to a data-scape that is unprecedented in its granular mapping of real-world consumption and transportation habits.

China's O2O explosion gave its companies tremendous data on the offline lives of their users: the what, where, and when of their meals, massages, and day-to-day activities. Digital payments cracked open the black box of real-world consumer purchases, giving these companies a precise, real-time data map of consumer behavior. Peer-to-peer transactions added a new layer of social data atop those economic transactions. The country's bike-sharing revolution has carpeted its cities in IoT transportation devices that color in the texture of urban life. They trace tens of millions of commutes, trips to the store, rides home, and first dates, dwarfing companies like Uber and Lyft in both quantity and granularity of data.

The numbers for these categories lay bare the China-U.S. gap in these key industries. Recent estimates have Chinese companies outstripping U.S. competitors ten to one in quantity of food deliveries and fifty to one in spending on mobile payments. China's e-commerce purchases are roughly double the U.S. totals, and the gap is only growing. Data on total trips through ride-hailing apps is somewhat scarce, but during the height of competition between Uber and Didi, self-reported numbers from the two companies had Didi's rides in China at four times the total of Uber's global rides. When it comes to rides on shared bikes, China is outpacing the United States at an astounding ratio of three hundred to one.

That has already helped China's juggernauts make up ground

on their American counterparts in both revenue and market caps. In the age of AI implementation, the impact of these divergent data ecosystems will be far more profound. It will shape what industries AI startups will disrupt in each country and what intractable problems they will solve.

But building an AI-driven economy requires more than just gladiator entrepreneurs and abundant data. It also takes an army of trained AI engineers and a government eager to embrace the power of this transformative technology. These two factors — AI expertise and government support — are the final pieces of the AI puzzle. When put in place, they will complete our analysis of the competitive balance between the world's two superpowers in the defining technology of the twenty-first century.

4

A TALE OF TWO COUNTRIES

Back in 1999, Chinese researchers were still in the dark when it came to studying artificial intelligence — literally. Allow me to explain.

That year, I visited the University of Science and Technology of China to give a lecture about our work on speech and image recognition at Microsoft Research. The university was one of the best engineering schools in the country, but it was located in the southern city of Hefei (pronounced "Huh-faye"), a remote backwater compared with Beijing.

On the night of the lecture, students crammed into the auditorium, and those who couldn't get a ticket pressed up against the windows, hoping to catch some of the lecture through the glass. Interest was so intense that I eventually asked the organizers to allow students to fill up the aisles and even sit on the stage around me. They listened intently as I laid out the fundamentals of speech recognition, speech synthesis, 3-D graphics, and computer vision. They scribbled down notes and peppered me with questions about underlying principles and practical applications. China clearly lagged behind the United States by more than a decade in AI research, but these students were like sponges for any knowledge from the outside world. The excitement in the room was palpable.

The lecture ran long, and it was already dark as I left the auditorium and headed toward the university's main gate. Student dorms lined both sides of the road, but the campus was still and the street was empty. And then, suddenly, it wasn't. As if on cue, long lines of students began pouring out of the dormitories all around me and walking out into the street. I stood there baffled, watching what looked like a slow-motion fire drill, all of it conducted in total silence.

It wasn't until they sat down on the curb and opened up their textbooks that I realized what was going on: the dormitories turned off all their lights at 11 p.m. sharp, and so most of the student body headed outside to continue their studies by streetlight. I looked on as hundreds of China's brightest young engineering minds huddled in the soft yellow glow. I didn't know it at the time, but the future founder of one of China's most important AI companies was there, squeezing in an extra couple of hours of studying in the dark Hefei night.

Many of the textbooks these students read were outdated or poorly translated. But they were the best the students could get their hands on, and these young scholars were going to wring them for every drop of knowledge they contained. Internet access at the school was a scarce commodity, and studying abroad was possible only if the students earned a full scholarship. The dog-eared pages of these textbooks and the occasional lecture from a visiting scholar were the only window they had into the state of global AI research.

Oh, how things have changed.

THE STUFF OF AN AI SUPERPOWER

As I laid out earlier, creating an AI superpower for the twenty-first century requires four main building blocks: abundant data, tenacious entrepreneurs, well-trained AI scientists, and a supportive policy environment. We've already seen how China's gladiatorial startup ecosystem trained a generation of the world's most street-smart entrepreneurs, and how China's alternate internet universe created the world's richest data ecosystem.

This chapter assesses the balance of power in the two remaining ingredients — AI expertise and government support. I believe that in the age of AI implementation, Silicon Valley's edge in elite expertise isn't all it's cracked up to be. And in the crucial realm of government support, China's techno-utilitarian political culture will pave the way for faster deployment of game-changing technologies.

As artificial intelligence filters into the broader economy, this era will reward the *quantity* of solid AI engineers over the *quality* of elite researchers. Real economic strength in the age of AI implementation won't come just from a handful of elite scientists who push the boundaries of research. It will come from an army of well-trained engineers who team up with entrepreneurs to turn those discoveries into game-changing companies.

China is training just such an army. In the two decades since my lecture in Hefei, China's artificial intelligence community has largely closed the gap with the United States. While America still dominates when it comes to superstar researchers, Chinese companies and research institutions have filled their ranks with the kind of well-trained engineers that can power this era of AI deployment. It has done that by marrying the extraordinary hunger for knowledge that I witnessed in Hefei with an explosion in access to cutting-edge global research. Chinese students of AI are no longer straining in the dark to read outdated textbooks. They're taking advantage of AI's open research culture to absorb knowledge straight from the source and in real time. That means dissecting the latest online academic publications, debating the approaches of top AI scientists in WeChat groups, and streaming their lectures on smartphones.

This rich connectivity allows China's AI community to play intellectual catch-up at the elite level, training a generation of hungry Chinese researchers who now contribute to the field at a high level. It also empowers Chinese startups to apply cutting-edge, open source algorithms to practical AI products: autonomous drones, pay-with-your-face systems, and intelligent home appliances.

Those startups are now scrapping for a slice of an AI landscape increasingly dominated by a handful of major players: the so-called Seven Giants of the AI age, which include Google, Facebook, Amazon, Microsoft, Baidu, Alibaba, and Tencent. These corporate juggernauts are almost evenly split between the United States and China, and they're making bold plays to dominate the AI economy. They're using billions of dollars in cash and dizzying stockpiles of data to gobble up available AI talent. They're also working to construct the "power grids" for the AI age: privately controlled computing networks that distribute machine learning across the economy, with the corporate giants acting as "utilities." It's a worrisome phenomenon for those who value an open AI ecosystem and also poses a potential stumbling block to China's rise as an AI superpower.

But bringing AI's power to bear on the broader economy can't be done by private companies alone — it requires an accommodating policy environment and can be accelerated by direct government support. As you recall, soon after Ke Jie's loss to AlphaGo, the Chinese central government released a sweeping blueprint for Chinese leadership in AI. Like the "mass innovation and mass entrepreneurship" campaign, China's AI plan is turbocharging growth through a flood of new funding, including subsidies for AI startups and generous government contracts to accelerate adoption.

The plan has also shifted incentives for policy innovation around AI. Ambitious mayors across China are scrambling to turn their cities into showcases for new AI applications. They're plotting driverless trucking routes, installing facial recognition systems on public transportation, and hooking traffic grids into "city brains" that optimize flows.

Behind these efforts lies a core difference in American and Chinese political culture: while America's combative political system aggressively punishes missteps or waste in funding technological upgrades, China's techno-utilitarian approach rewards proactive investment and adoption. Neither system can claim objective moral superiority, and the United States' long track record of both personal freedom and technological achievement is unparalleled in the modern era. But I believe that in the age of AI implementation the Chinese approach will have the impact of accelerating deployment, generating more data, and planting the seeds of further growth. It's a self-perpetuating cycle, one that runs on a peculiar alchemy of digital data, entrepreneurial grit, hard-earned expertise, and political will. To see where the two AI superpowers stand, we must first understand the source of that expertise.

When Enrico Fermi stepped onto the deck of the RMS *Franconia II* in 1938, he changed the global balance of power. Fermi had just received the Nobel Prize in physics in Stockholm, but instead of returning home to Benito Mussolini's Italy, Fermi and his family sailed for New York. They made the journey to escape Italy's racial laws, which barred Jews or Africans from holding many jobs or marrying Italians. Fermi's wife, Laura, was Jewish, and he decided to move the family halfway across the world rather than live under the antisemitism that was sweeping Europe.

It was a personal decision with earthshaking consequences. After arriving in the United States, Fermi learned of the discovery of nuclear fission by scientists in Nazi Germany and quickly set to work exploring the phenomenon. He created the world's first self-sustaining nuclear reaction underneath a set of bleachers at the University of Chicago and played an indispensable role in the Manhattan Project. This top-secret project was the largest industrial undertaking the world had ever seen, and it culminated in the development of the world's first nuclear weapons for the U.S. military. Those bombs put an end to World War II in the Pacific and laid the groundwork for the nuclear world order.

Fermi and the Manhattan Project embodied an age of discovery that rewarded quality over quantity in expertise. In nuclear physics, the 1930s and 1940s were an age of fundamental breakthroughs, and when it came to making those breakthroughs, one Enrico Fermi was worth thousands of less brilliant physicists. American leadership in this era was built in large part on attracting geniuses like Fermi: men and women who could singlehandedly tip the scales of scientific power.

But not every technological revolution follows this pattern. Often, once a fundamental breakthrough has been achieved, the center of gravity quickly shifts from a handful of elite researchers to an army of tinkerers — engineers with just enough expertise to apply the technology to different problems. This is particularly true when the payoff of a breakthrough is diffused throughout society rather than concentrated in a few labs or weapons systems.

Mass electrification exemplified this process. Following Thomas Edison's harnessing of electricity, the field rapidly shifted from invention to implementation. Thousands of engineers began tinkering with electricity, using it to power new devices and reorganize industrial processes. Those tinkerers didn't have to break new ground like Edison. They just had to know enough about how electricity worked to turn its power into useful and profitable machines.

Our present phase of AI implementation fits this latter model. A constant stream of headlines about the latest task tackled by AI gives us the mistaken sense that we are living through an age of discovery, a time when the Enrico Fermis of the world determine the balance of power. In reality, we are witnessing the application of one fundamental breakthrough — deep learning and related techniques — to many different problems. That's a process that requires well-trained AI scientists, the tinkerers of this age. Today, those tinkerers are putting AI's superhuman powers of pattern recognition to use making loans, driving cars, translating text, playing Go, and powering your Amazon Alexa.

Deep-learning pioneers like Geoffrey Hinton, Yann LeCun, and Yoshua Bengio — the Enrico Fermis of AI — continue to push the boundaries of artificial intelligence. And they may yet produce another game-changing breakthrough, one that scrambles the global technological pecking order. But in the meantime, the real action today is with the tinkerers.

INTELLIGENCE SHARING

And for this technological revolution, the tinkerers have an added advantage: real-time access to the work of leading pioneers. During the Industrial Revolution, national borders and language barriers meant that new industrial breakthroughs remained bottled up in their country of origin, England. America's cultural proximity and loose intellectual property laws helped it pilfer some key inventions, but there remained a substantial lag between the innovator and the imitator.

Not so today. When asked how far China lags behind Silicon Valley in artificial intelligence research, some Chinese entrepreneurs jokingly answer "sixteen hours" — the time difference between California and Beijing. America may be home to the top researchers, but much of their work and insight is instantaneously available to anyone with an internet connection and a grounding in AI fundamentals. Facilitating this knowledge transfer are two defining traits of the AI research community: openness and speed.

Artificial intelligence researchers tend to be quite open about publishing their algorithms, data, and results. That openness grew out of the common goal of advancing the field and also from the desire for objective metrics in competitions. In many physical sciences, experiments cannot be fully replicated from one lab to the next — minute variations in technique or environment can greatly affect results. But AI experiments are perfectly replicable, and algorithms are directly comparable. They simply require those algorithms to be trained and tested on identical data sets. International competitions frequently pit different computer vision or speech recognition teams against each other, with the competitors opening their work to scrutiny by other researchers.

The speed of improvements in AI also drives researchers to instantly share their results. Many AI scientists aren't trying to make fundamental breakthroughs on the scale of deep learning, but they are constantly making marginal improvements to the best algorithms. Those improvements regularly set new records for accuracy on tasks like speech recognition or visual identification. Researchers compete on the basis of these records — not on new products or revenue numbers — and when one sets a new record, he or she wants to be recognized and receive credit for the achievement. But given the rapid pace of improvements, many researchers fear that if they wait to publish in a journal, their record will already have been eclipsed and their moment at the cutting edge will go undocumented. So instead of sitting on that research, they opt for instant publication on websites like www.arxiv.org, an online repository of scientific papers. The site lets researchers instantly time-stamp their research, planting a stake in the ground to mark the "when and what" of their algorithmic achievements.

In the post-AlphaGo world, Chinese students, researchers, and engineers are among the most voracious readers of arxiv.org. They trawl the site for new techniques, soaking up everything the world's top researchers have to offer. Alongside these academic publications, Chinese AI students also stream, translate, and subtitle lectures from leading AI scientists like Yann LeCun, Stanford's Sebastian Thrun, and Andrew Ng. After decades spent studying outdated textbooks in the dark, these researchers revel in this instant connectivity to global research trends.

On WeChat, China's AI community coalesces in giant group chats and multimedia platforms to chew over what's new in AI. Thirteen new media companies have sprung up just to cover the sector, offering industry news, expert analysis, and open-ended dialogue. These AI-focused outlets boast over a million registered users, and half of them have taken on venture funding that values them at more than $10 million each. For more academic discussions, I'm part of the five-hundred-member "Weekly Paper Discussion Group," just one of the dozens of WeChat groups that come together to dissect a new AI research publication each week. The chat group buzzes with hundreds of messages per day: earnest questions about this week's paper, screen shots of the members' latest algorithmic achievements, and, of course, plenty of animated emojis.

But Chinese AI practitioners aren't just passive recipients of wisdom spilling forth from the Western world. They're now giving back to that research ecosystem at an accelerating rate.

CONFERENCE CONFLICTS

The Association for the Advancement of Artificial Intelligence had a problem. The storied organization had been putting on one of the world's most important AI conferences for three decades, but in 2017 they were in danger of hosting a dud event.

Why? The conference dates conflicted with Chinese New Year.

A few years earlier, this wouldn't have been a problem. Historically, American, British, and Canadian scholars have dominated the proceedings, with just a handful of Chinese researchers presenting papers. But the 2017 conference had accepted an almost equal number of papers from researchers in China and the United States, and it was in danger of losing half of that equation to their culture's most important holiday.

"Nobody would have put AAAI on Christmas day," the group's president told the *Atlantic*. "Our organization had to almost turn on a dime and change the conference venue to hold it a week later."

Chinese AI contributions have occurred at all levels, ranging from marginal tweaks of existing models to the introduction of world-class new approaches to neural network construction. A look at citations in academic research reveals the growing clout of Chinese researchers. One study by Sinovation Ventures examined citations in the top one hundred AI journals and conferences from 2006 to 2015; it found that the percentage of papers by authors with Chinese names nearly doubled from 23.2 percent to 42.8 percent during that time. That total includes some authors with Chinese names who work abroad — for example, Chinese American researchers who haven't adopted an anglicized name. But a survey of the authors' research institutions found the great majority of them to be working in China.

A recent tally of citations at global research institutions confirmed the trend. That ranking of the one hundred most-cited research institutions on AI from 2012 to 2016 showed China ranking second only to the United States. Among the elite institutions, Tsinghua University even outnumbered places like Stanford University in total AI citations. These studies largely captured the pre-AlphaGo era, before China pushed even more researchers into the field. In the coming years, a whole new wave of young Ph.D. students will bring Chinese AI research to a new level.

And these contributions haven't just been about piling up papers and citations. Researchers in the country have produced some of the most important advances in neural networks and computer vision since the arrival of deep learning. Many of these researchers emerged out of Microsoft Research China, an institution that I founded in 1998. Later renamed Microsoft Research Asia, it went on to train over five thousand AI researchers, including top executives at Baidu, Alibaba, Tencent, Lenovo, and Huawei.

In 2015, a team from Microsoft Research Asia blew the competition out of the water at the global image-recognition competition, ImageNet. The team's breakthrough algorithm was called ResNet, and it identified and classified objects from 100,000 photographs into 1,000 different categories with an error rate of just 3.5 percent. Two years later, when Google's DeepMind built AlphaGo Zero — the self-taught successor to AlphaGo — they used ResNet as one of its core technological building blocks.

The Chinese researchers behind ResNet didn't stay at Microsoft for long. Of the four authors of the ResNet paper, one joined Yann LeCun's research team at Facebook, but the other three have founded and joined AI startups in China. One of those startups, Face++, has quickly turned into a world leader in face- and image-recognition technology. At the 2017 COCO image-recognition competition, the Face++ team took first place in three of the four most important categories, beating out the top teams from Google, Microsoft, and Facebook.

To some observers in the West, these research achievements fly in the face of deeply held beliefs about the nature of knowledge and research across political systems. Shouldn't Chinese controls on the internet hobble the ability of Chinese researchers to break new ground globally? There are valid critiques of China's system of governance, ones that weigh heavily on public debate and research in the social sciences. But when it comes to research in the hard sciences, these issues are not nearly as limiting as many outsiders presume. Artificial intelligence doesn't touch on sensitive political questions, and China's AI scientists are essentially as free as their American counterparts to construct cutting-edge algorithms or build profitable AI applications.

But don't take it from me. At a 2017 conference on artificial intelligence and global security, former Google CEO Eric Schmidt warned participants against complacency when it came to Chinese AI capabilities. Predicting that China would match American AI capabilities in five years, Schmidt was blunt in his assessment: "Trust me, these Chinese people are good. . . . If you have any kind of prejudice or concern that somehow their system and their educational system is not going to produce the kind of people that I'm talking about, you're wrong."

But while the global AI research community has blossomed into a fluid and open system, one component of that ecosystem remains more closed off: big corporate research labs. Academic researchers may rush to share their work with the world, but public technology companies have a fiduciary responsibility to maximize profits for their shareholders. That usually means less publishing and more proprietary technology.

Of the hundreds of companies pouring resources into AI research, let's return to the seven that have emerged as the new giants of corporate AI research — Google, Facebook, Amazon, Microsoft, Baidu, Alibaba, and Tencent. These Seven Giants have, in effect, morphed into what nations were fifty years ago — that is, large and relatively closed-off systems that concentrate talent and resources on breakthroughs that will mostly remain "in house."

The seals around corporate research are never airtight: team members leave to found their own AI startups, and some groups like Microsoft Research, Facebook AI Research, and DeepMind still publish articles on their most meaningful contributions. But broadly speaking, if one of these companies makes a unique breakthrough — a trade secret that could generate massive profits for that company alone — it will do its best to keep a lid on it and will try to extract maximum value before the word gets out.

A groundbreaking discovery occurring within one of these closed systems poses the greatest threat to the world's open AI ecosystem. It also threatens to stymie China in its goal of becoming a global leader in AI. The way things stand today, China already has the edge in entrepreneurship, data, and government support, and it's rapidly catching up to the United States in expertise. If the technological status quo holds for the coming years, an array of Chinese AI startups will begin fanning out across different industries. They will leverage deep learning and other machine-learning technologies to disrupt dozens of sectors and reap the rewards of transforming the economy.

But if the next breakthrough on the scale of deep learning occurs soon, and it happens within a hermetically sealed corporate environment, all bets are off. It could give one company an insurmountable advantage over the other Seven Giants and return us to an age of discovery in which elite expertise tips the balance of power in favor of the United States.

To be clear, I believe the odds are slightly against such a breakthrough coming out of the corporate behemoths in the coming years. Deep learning marked the largest leap forward in the past fifty years, and advances on this scale rarely come more than once every few decades. Even if such a breakthrough does occur, it's more likely to emerge out of the open environment of academia. Right now, the corporate giants are pouring unprecedented resources into squeezing deep learning for all it's worth. That means lots of fine-tuning of deep-learning algorithms and only a small percentage of truly open-ended research in pursuit of the next paradigm-shifting breakthrough.

Meanwhile, academics find themselves unable to compete with industry in practical applications of deep learning because of the requirements for massive amounts of data and computing power. So instead, many academic researchers are following Geoffrey Hinton's exhortation to move on and focus on inventing "the next deep learning," a fundamentally new approach to AI problems that could change the game. That type of open-ended research is the kind most likely to stumble onto the next breakthrough and then publish it for all the world to learn from.

GOOGLE VERSUS THE REST

But if the next deep learning *is* destined to be discovered in the corporate world, Google has the best shot at it. Among the Seven AI Giants, Google — more precisely, its parent company, Alphabet, which owns DeepMind and its self-driving subsidiary Waymo — stands head and shoulders above the rest. It was one of the earliest companies to see the potential in deep learning and has devoted more resources to harnessing it than any other company.

In terms of funding, Google dwarfs even its own government: U.S.

federal funding for math and computer science research amounts to less than half of Google's own R&D budget. That spending spree has bought Alphabet an outsized share of the world's brightest AI minds. Of the top one hundred AI researchers and engineers, around half are already working for Google.

The other half are distributed among the remaining Seven Giants, academia, and a handful of smaller startups. Microsoft and Facebook have soaked up substantial portions of this group, with Facebook bringing on superstar researchers like Yann LeCun. Of the Chinese giants, Baidu went into deep-learning research earliest — even trying to acquire Geoffrey Hinton's startup in 2013 before being outbid by Google — and scored a major coup in 2014 when it recruited Andrew Ng to head up its Silicon Valley AI Lab. Within a year, that hire was showing outstanding results. By 2015, Baidu's AI algorithms had exceeded human abilities at Chinese speech recognition. It was a great accomplishment, but one that went largely unnoticed in the United States. In fact, when Microsoft reached the same milestone a year later for English, the company dubbed it a "historic achievement." Ng left Baidu in 2017 to create his own AI investment fund, but the time he spent at the company both testified to Baidu's ambitions and strengthened its reputation for research.

Alibaba and Tencent were relative latecomers to the AI talent race, but they have the cash and data on hand to attract top talent. With WeChat serving as the all-in-one super-app of the world's largest internet market, Tencent possesses perhaps the single richest data ecosystem of all the giants. That is now helping Tencent to attract and empower top-flight AI researchers. In 2017, Tencent opened an AI research institute in Seattle and immediately began poaching Microsoft researchers to staff it.

Alibaba has followed suit with plans to open a global network of research labs, including in Silicon Valley and Seattle. Thus far, Tencent and Alibaba have yet to publicly demonstrate the results of this research, opting instead for more product-driven applications. Alibaba has taken the lead on "City Brains": massive AI-driven networks that optimize city services by drawing on data from video cameras, social media, public transit, and location-based apps. Working with the city government in its hometown of Hangzhou, Alibaba is using advanced object-recognition and predictive transit algorithms to constantly tweak the patterns for red lights and alert emergency services to traffic accidents. The trial has increased traffic speeds by 10 percent in some areas, and Alibaba is now preparing to bring the service to other cities.

While Google may have jumped off to a massive head start in the arms race for elite AI talent, that by no means guarantees victory. As discussed, fundamental breakthroughs are few and far between, and paradigm-shifting discoveries often emerge from unexpected places. Deep learning came out of a small network of idiosyncratic researchers obsessed with an approach to machine learning that had been dismissed by mainstream researchers. If the next deep learning is out there somewhere, it could be hiding on any number of university campuses or in corporate labs, and there's no guessing when or where it will show its face. While the world waits for the lottery of scientific discovery to produce a new breakthrough, we remain entrenched in our current era of AI implementation.

POWER GRIDS VERSUS AI BATTERIES

But the giants aren't just competing against one another in a race for the next deep learning. They're also in a more immediate race against the small AI startups that want to use machine learning to revolutionize specific industries. It's a contest between two approaches to distributing the "electricity" of AI across the economy: the "grid" approach of the Seven Giants versus the "battery" approach of the startups. How that race plays out will determine the nature of the AI business landscape — monopoly, oligopoly, or freewheeling competition among hundreds of companies.

The "grid" approach is trying to commoditize AI. It aims to turn the power of machine learning into a standardized service that can be purchased by any company — or even be given away for free for academic or personal use — and accessed via cloud computing platforms. In this model, cloud computing platforms act as the grid, performing complex machine-learning optimizations on whatever data

problems users require. The companies behind these platforms —
Google, Alibaba, and Amazon — act as the utility companies, managing the grid and collecting the fees.

Hooking into that grid would allow traditional companies with large data sets to easily tap into AI's optimization powers without having to remake their entire business around it. Google's Tensor-Flow, an open-source software ecosystem for building deep learning-models, offers an early version of this but still requires some AI expertise to operate. The goal of the grid approach is to both lower that expertise threshold and increase the functionality of these cloud-based AI platforms. Making use of machine learning is nowhere near as simple as plugging an electric appliance into the wall — and it may never be — but the AI giants hope to push things in that direction and then reap the rewards of generating the "power" and operating the "grid."

AI startups are taking the opposite approach. Instead of waiting for this grid to take shape, startups are building highly specific "battery-powered" AI products for each use-case. These startups are banking on depth rather than breadth. Instead of supplying general-purpose machine-learning capabilities, they build new products and train algorithms for specific tasks, including medical diagnosis, mortgage lending, and autonomous drones.

They are betting that traditional businesses won't be able to simply plug the nitty-gritty details of their daily operations into an all-purpose grid. Instead of helping those companies access AI, these startups want to disrupt them using AI. They aim to build AI-first companies from the ground up, creating a new roster of industry champions for the AI age.

It's far too early to pick a winner between the grid and battery approaches. While giants like Google steadily spread their tentacles outward, startups in China and the United States are racing to claim virgin territory and fortify themselves against incursions by the Seven Giants. How that scramble for territory shakes out will determine the shape of our new economic landscape. It could concentrate astronomical profits in the hands of the Seven Giants — the super-utilities of the AI age — or diffuse those profits out across thousands of vibrant new companies.

One underdiscussed area of AI competition—among the AI giants, startups, and the two countries—is in computer chips, also known as semiconductors. High-performance chips are the unsexy, and often unsung, heroes of each computing revolution. They are at the literal core of our desktops, laptops, smartphones, and tablets, but for that reason they remain largely hidden to the end user. But from an economic and security perspective, building those chips is a very big deal: the markets tend toward lucrative monopolies, and security vulnerabilities are best spotted by those who work directly with the hardware.

Each era of computing requires different kinds of chips. When desktops reigned supreme, chipmakers sought to maximize processing speed and graphics on a high-resolution screen, with far less concern about power consumption. (Desktops were, after all, always plugged in.) Intel mastered the design of these chips and made billions in the process. But with the advent of smartphones, demand shifted toward more efficient uses of power, and Qualcomm, whose chips were based on designs by the British firm ARM, took the throne as the undisputed chip king.

Now, as traditional computing programs are displaced by the operation of AI algorithms, requirements are once again shifting. Machine learning demands the rapid-fire execution of complex mathematical formulas, something for which neither Intel's nor Qualcomm's chips are built. Into the void stepped Nvidia, a chipmaker that had previously excelled at graphics processing for video games. The math behind graphics processing aligned well with the requirements for AI, and Nvidia became the go-to player in the chip market. Between 2016 and early 2018, the company's stock price multiplied by a factor of ten.

These chips are central to everything from facial recognition to self-driving cars, and that has set off a race to build the next-generation AI chip. Google and Microsoft—companies that had long avoided building their own chips—have jumped into the fray, alongside Intel, Qualcomm, and a batch of well-funded Silicon Valley chip startups. Facebook has partnered with Intel to test-drive its first foray into AI-specific chips.

But for the first time, much of the action in this space is taking place in China. The Chinese government has for many years — decades, even — tried to build up indigenous chip capabilities. But constructing a high-performance chip is an extremely complex and expertise-intensive process, one that has so far remained impervious to several government-sponsored projects. For the last three decades, it's been private Silicon Valley firms that have cashed in on chip development.

Chinese leaders and a raft of chip startups are hoping that this time is different. The Chinese Ministry of Science and Technology is doling out large sums of money, naming as a specific goal the construction of a chip with performance and energy efficiency twenty times better than one of Nvidia's current offerings. Chinese chip startups like Horizon Robotics, Bitmain, and Cambricon Technologies are flush with investment capital and working on products tailor-made for self-driving cars or other AI use-cases. The country's edge in data will also feed into chip development, offering hardware makers a feast of examples on which to test their products.

On balance, Silicon Valley remains the clear leader in AI chip development. But it's a lead that the Chinese government and the country's venture-capital community are trying their best to erase. That's because when economic disruption occurs on the scale promised by artificial intelligence, it isn't just a business question — it's also a major political question.

A TALE OF TWO AI PLANS

On October 12, 2016, President Barack Obama's White House released a long-brewing plan for how the United States can harness the power of artificial intelligence. The document detailed the transformation AI is set to bring to the economy and laid out steps to seize that opportunity: increasing funding for research, stepping up civilian-military cooperation, and making investments to mitigate social disruptions. It offered a decent summary of changes on the horizon and some commonsense prescriptions for adaptation.

But the report — issued by the most powerful political office in the United States — had about the same impact as a wonkish policy paper from an academic think tank. Released the same week as Donald Trump's infamous *Access Hollywood* videotape, the White House report barely registered in the American news cycle. It did not spark a national surge in interest about AI. It did not lead to a flood of new VC investments and government funding for AI start-ups. And it didn't galvanize mayors or governors to adopt AI-friendly policies. In fact, when President Trump took office just three months after the report's debut, he proposed *cutting* funding for AI research at the National Science Foundation.

The limp response to the Obama report made for a stark contrast to the shockwaves generated by the Chinese government's own AI plan. Like past Chinese government documents on technology, it was plain in its language but momentous in its impact. Published in July 2017, the Chinese State Council's "Development Plan for a New Generation of Artificial Intelligence" shared many of the same predictions and recommendations as the White House plan. It also spelled out hundreds of industry-specific applications of AI and laid down signposts for China's progress toward becoming an AI super-power. It called for China to reach the top tier of AI economies by 2020, achieve major new breakthroughs by 2025, and become the global leader in AI by 2030.

If AlphaGo was China's Sputnik Moment, the government's AI plan was like President John F. Kennedy's landmark speech calling for America to land a man on the moon. The report lacked Kennedy's soaring rhetoric, but it set off a similar national mobilization, an all-hands-on-deck approach to national innovation.

BETTING ON AI

China's AI plan originated at the highest levels of the central government, but China's ambitious mayors are where the real action takes place. Following the release of the State Council's plan, local officials angling for promotion threw themselves into the goal of turning their cities into hubs for AI development. They offered subsidies for research, directed venture-capital "guiding funds" toward

AI, purchased the products and services of local AI startups, and set up dozens of special development zones and incubators.

We can see the intricacy of these support policies by zooming in on one city, Nanjing. The capital of Jiangsu province on China's eastern seaboard, Nanjing is not among the top tier of Chinese cities for startups — those honors go to Beijing, Shenzhen, and Hangzhou. But in a bid to transform Nanjing into an AI hotspot, the city government is pouring vast sums of money and policy resources into attracting AI companies and top talent.

Between 2017 and 2020, the Nanjing Economic and Technological Development Zone plans to put at least 3 billion RMB (around $450 million) into AI development. That money will go toward a dizzying array of AI subsidies and perks, including investments of up to 15 million RMB in local companies, grants of 1 million RMB per company to attract talent, rebates on research expenses of up to 5 million RMB, creation of an AI training institute, government contracts for facial recognition and autonomous robot technology, simplified procedures for registering a company, seed funding and office space for military veterans, free company shuttles, coveted spots at local schools for the children of company executives, and special apartments for employees of AI startups.

And that is all in just one city. Nanjing's population of 7 million ranks just tenth in China, a country with a hundred cities of more than a million people. This blizzard of government incentives is going on across many of those cities right now, all competing to attract, fund, and empower AI companies. It's a process of government-accelerated technological development that I've witnessed twice in the past decade. Between 2007 and 2017, China went from having zero high-speed rail lines to having more miles of high-speed rail operational than the rest of the world combined. During the "mass innovation and mass entrepreneurship" campaign that began in 2015, a similar flurry of incentives created 6,600 new startup incubators and shifted the national culture around technology startups.

Of course, it's too early to know the exact results of China's AI campaign, but if Chinese history is any guide, it is likely to be somewhat inefficient but extremely effective. The sheer scope of financing and speed of deployment almost guarantees that there will be inefficiencies. Government bureaucracies cannot rapidly deploy billions of dollars in investments and subsidies without some amount of waste. There will be dorms for AI employees that will never be inhabited, and investments in startups that will never get off the ground. There will be traditional technology companies that merely rebrand themselves as "AI companies" to rake in subsidies, and AI equipment purchases that simply gather dust in government offices.

But that's a risk these Chinese government officials are willing to take, a loss they're willing to absorb in pursuit of a larger goal: brute-forcing the economic and technological upgrading of their cities. The potential upside of that transformation is large enough to warrant making expensive bets on the next big thing. And if the bet doesn't pan out, the mayors won't be endlessly pilloried by their opponents for attempting to act on the central government's wishes.

Contrast that with the political firestorm following big bets gone bad in the United States. After the 2008 financial crisis, President Obama's stimulus program included plans for government loan guarantees on promising renewable energy projects. It was a program designed to stimulate a stagnant economy but also to facilitate a broader economic and environmental shift toward green energy.

One of the recipients of those loan guarantees was Solyndra, a California solar panel company that initially looked promising but then went bankrupt in 2011. President Obama's critics quickly turned that failure into one of the most potent political bludgeons of the 2012 presidential election. They hammered the president with millions of dollars in attack ads, criticizing the "wasteful" spending as a symptom of "crony capitalism" and "venture socialism." Never mind that, on the whole, the loan guarantee program is projected to *earn* money for the federal government — one high-profile failure was enough to tar the entire enterprise of technological upgrading.

Obama survived the negative onslaught to win another term, but the lessons for American politicians were clear: using government funding to invest in economic and technological upgrades is a risky business. Successes are often ignored, and every misfire becomes fodder for attack ads. It's far safer to stay out of the messy business of upgrading an economy.

That same divide in political cultures applies to creating a supportive policy environment for AI development. For the past thirty years, Chinese leaders have practiced a kind of techno-utilitarianism, leveraging technological upgrades to maximize broader social good while accepting that there will be downsides for certain individuals or industries. It, like all political structures, is a highly imperfect system. Top-down government mandates to expand investment and production can also send the pendulum of public investment swinging too far in a given direction. In recent years, this has led to massive gluts of supply and unsustainable debt loads in Chinese industries ranging from solar panels to steel. But when national leaders correctly channel those mandates toward new technologies that can lead to seismic economic shifts, the techno-utilitarian approach can have huge upsides.

Self-driving cars make for a good example of this balancing act. In 2016, the United States lost forty thousand people to traffic accidents. That annual death toll is equivalent to the 9/11 terrorist attacks occurring once every month from January through November, and twice in December. The World Health Organization estimates that there are around 260,000 annual road fatalities in China and 1.25 million around the globe.

Autonomous vehicles are on the path to eventually being far safer than human-driven vehicles, and widespread deployment of the technology will dramatically decrease these fatalities. It will also lead to huge increases in efficiency of transportation and logistics networks, gains that will echo throughout the entire economy.

But alongside the lives saved and productivity gained, there will be other instances in which jobs or even lives are lost due to the very same technology. For starters, taxi, truck, bus, and delivery drivers will be largely out of luck in a self-driving world. There will also inevitably be malfunctions in autonomous vehicles that cause crashes. There will be circumstances that force an autonomous vehicle to make agonizing ethical decisions, like whether to veer right and have a 55 percent chance of killing two people or veer left and have a 100 percent chance of killing one person.

Every one of these downside risks presents thorny ethical questions. How should we balance the livelihoods of millions of truck drivers against the billions of dollars and millions of hours saved by autonomous vehicles? What should a self-driving car "optimize for" in situations where it is forced to choose which car to crash into? How should an autonomous vehicle's algorithm weigh the life of its owner? Should your self-driving car sacrifice your own life to save the lives of three other people?

These are the questions that keep ethicists up at night. They're also questions that could hold up the legislation needed for autonomous-vehicle deployment and tie up AI companies in years of lawsuits. They may well lead American politicians, ever fearful of interest groups and attack ads, to pump the brakes on widespread self-driving vehicle deployment. We've already seen early signs of this happening, with unions representing truck drivers successfully lobbying Congress in 2017 to exclude trucks from legislation aimed at speeding up autonomous-vehicle deployment.

I believe the Chinese government will see these difficult concerns as important topics to explore but not as a reason to delay the implementation of technology that will save tens if not hundreds of thousands of lives in the not-too-distant future. For better or worse — and I recognize that most Americans may not embrace this view — Chinese political culture doesn't carry the American expectation of reaching a moral consensus on each of the above questions. Promotion of a broader social good — the long-term payoff in lives saved — is a good enough reason to begin implementation, with outlier cases and legal intricacies to be dealt with in due time. Again, this is not a call for the United States and Europe to mimic the techno-utilitarian approach utilized in China — every country should decide on its own approach based on its own cultural values. But it's important to understand the Chinese approach and the implications it holds for the pace and path of AI development.

Accelerating that deployment will feature the same scramble by local government officials to stand out on AI. Along with competing to attract AI companies through subsidies, these mayors and provincial governors will compete to be the first to implement high-profile AI projects, such as AI-assisted doctors at public hospitals or autonomous trucking routes and "city brains" that optimize urban traffic grids. They can pursue these projects for both the political points scored and the broad social upside, spending less time obsessing over the downside risks that would scare away risk-sensitive American politicians.

This is not an ethical judgment on either of these two systems. Utilitarian government systems and rights-based approaches both have their blind spots and downsides. America's openness to immigration and emphasis on individual rights has long helped it attract some of the brightest minds from around the world — people like Enrico Fermi, Albert Einstein, and many leading AI scientists today. China's top-down approach to economic upgrades — and the eagerness of low-level officials to embrace each new central government mandate — can also lead to waste and debt if the target industries are not chosen well. But in this particular instance — building a society and economy prepared to harness the potential of AI — China's techno-utilitarian approach gives it a certain advantage. Its acceptance of risk allows the government to make big bets on game-changing technologies, and its approach to policy will encourage faster adoption of those technologies.

With these national strengths and weaknesses in mind, we can construct a timeline for AI deployment and look at how specific AI products and systems are set to change the world around us.

5

★

THE FOUR WAVES OF AI

The year 2017 marked the first time I heard Donald Trump speak fluent Chinese. During the U.S. president's first trip to China, he showed up on a big screen to welcome attendees at a major tech conference. He began his speech in English and then abruptly switched languages.

"AI is changing the world," he said, speaking in flawless Chinese but with typical Trump bluster. "And iFlyTek is really fantastic."

President Trump cannot, of course, speak Chinese. But AI is indeed changing the world, and Chinese companies like iFlyTek are leading the way. By training its algorithms on large data samples of President Trump's speeches, iFlyTek created a near-perfect digital model of his voice: intonation, pitch, and pattern of speech. It then recalibrated that vocal model for Mandarin Chinese, showing the world what Donald Trump might sound like if he grew up in a village outside Beijing. The movement of lips wasn't precisely synced to the Chinese words, but it was close enough to fool a casual viewer at first glance. President Obama got the same treatment from iFlyTek: a video of a real press conference but with his professorial style converted to perfect Mandarin.

"With the help of iFlyTek, I've learned Chinese," Obama intoned to the White House press corps. "I think my Chinese is better than Trump's. What do all of you think?"

iFlyTek might say the same to its own competitors. The Chinese company has racked up victories at a series of prestigious international AI competitions for speech recognition, speech synthesis, image recognition, and machine translation. Even in the company's "second language" of English, iFlyTek often beats teams from Google, DeepMind, Facebook, and IBM Watson in natural-language processing—that is, the ability of AI to decipher overall meaning rather than just words.

This success didn't come overnight. Back in 1999, when I started Microsoft Research Asia, my top-choice recruit was a brilliant young Ph.D. named Liu Qingfeng. He had been one of the students I saw filing out of the dorms to study under streetlights after my lecture in Hefei. Liu was both hardworking and creative in tackling research questions; he was one of China's most promising young researchers. But when we asked him to accept our scholarship offer and become a Microsoft intern and then an employee, he declined. He wanted to start his own AI speech company. I told him that he was a great young researcher but that China lagged too far behind American speech-recognition giants like Nuance, and there were fewer customers in China for this technology. To his credit, Liu ignored that advice and poured himself into building iFlyTek. Nearly twenty years and dozens of AI competition awards later, iFlyTek has far surpassed Nuance in capabilities and market cap, becoming the most valuable AI speech company in the world.

Combining iFlyTek's cutting-edge capabilities in speech recognition, translation, and synthesis will yield transformative AI products, including simultaneous translation earpieces that instantly convert your words and voice into any language. It's the kind of product that will soon revolutionize international travel, business, and culture, and unlock vast new stores of time, productivity, and creativity in the process.

THE WAVES

But it won't happen all at once. The complete AI revolution will take a little time and will ultimately wash over us in a series of four waves: internet AI, business AI, perception AI, and autonomous AI. Each of these waves harnesses AI's power in a different way, disrupting different sectors and weaving artificial intelligence deeper into the fabric of our daily lives.

The first two waves—internet AI and business AI—are already all around us, reshaping our digital and financial worlds in ways we can barely register. They are tightening internet companies' grip on our attention, replacing paralegals with algorithms, trading stocks, and diagnosing illnesses.

Perception AI is now digitizing our physical world, learning to recognize our faces, understand our requests, and "see" the world around us. This wave promises to revolutionize how we experience and interact with our world, blurring the lines between the digital and physical worlds. Autonomous AI will come last but will have the deepest impact on our lives. As self-driving cars take to the streets, autonomous drones take to the skies, and intelligent robots take over factories, they will transform everything from organic farming to highway driving and fast food.

These four waves all feed off different kinds of data, and each one presents a unique opportunity for the United States or China to seize the lead. We'll see that China is in a strong position to lead or co-lead in internet AI and perception AI, and will likely soon catch up with the United States in autonomous AI. Currently, business AI remains the only arena in which the United States maintains clear leadership.

Competition, however, won't play out in just these two countries. AI-driven services that are pioneered in the United States and China will then proliferate across billions of users around the globe, many of them in developing countries. Companies like Uber, Didi, Alibaba, and Amazon are already fiercely competing for these developing markets but adopting very different strategies. While Silicon Valley juggernauts are trying to conquer each new market with their own products, China's internet companies are instead investing in these countries' scrappy local startups as they try to fight off U.S. domination. It's a competition that's just getting started, and one that will have profound implications for the global economic landscape of the twenty-first century.

To understand how this coming competition will play out at home and abroad, we must first take a dive into each of the four waves of AI washing over our economies.

Internet AI already likely has a strong grip on your eyeballs, if not your wallet. Ever find yourself going down an endless rabbit hole of YouTube videos? Do video streaming sites have an uncanny knack for recommending that next video that you've just got to check out before you get back to work? Does Amazon seem to know what you'll want to buy before you do?

If so, then you have been the beneficiary (or victim, depending on how you value your time, privacy, and money) of internet AI. This first wave began almost fifteen years ago but finally went mainstream around 2012. Internet AI is largely about using AI algorithms as *recommendation engines:* systems that learn our personal preferences and then serve up content hand-picked for us.

The horsepower of these AI engines depends on the digital data they have access to, and there's currently no greater storehouse of this data than the major internet companies. But that data only becomes truly useful to algorithms once it has been labeled. In this case, "labeled" doesn't mean you have to actively rate the content or tag it with a keyword. Labels simply come from linking a piece of data with a specific outcome: bought versus didn't buy, clicked versus didn't click, watched until the end versus switched videos. Those labels — our purchases, likes, views, or lingering moments on a web page — are then used to train algorithms to recommend more content that we're likely to consume.

Average people experience this as the internet "getting better" — that is, at giving us what we want — and becoming more addictive as it goes. But it's also proof of the power of AI to learn about us through data and then optimize for what we desire. That optimization has been translated into massive increases in profits for established internet companies that make money off our clicks: the Googles, Baidus, Alibabas, and YouTubes of the world. Using internet AI, Alibaba can recommend products you're more likely to buy, Google can target you with ads you're more likely to click on, and YouTube can suggest videos that you're more likely to watch. Adopting those same methods in a different context, a company like Cambridge Analytica used Facebook data to better understand and target American voters during the 2016 presidential campaign. Revealingly, it was Robert Mercer, founder of Cambridge Analytica, who reportedly coined the famous phrase, "There's no data like more data."

ALGORITHMS AND EDITORS

First-wave AI has given birth to entirely new, AI-driven internet companies. China's leader in this category is Jinri Toutiao (meaning "today's headlines"; English name: "ByteDance"). Founded in 2012, Toutiao is sometimes called "the BuzzFeed of China" because both sites serve as hubs for timely viral stories. But virality is where the similarities stop. BuzzFeed is built on a staff of young editors with a knack for cooking up original content. Toutiao's "editors" are algorithms.

Toutiao's AI engines trawl the internet for content, using natural-language processing and computer vision to digest articles and videos from a vast network of partner sites and commissioned contributors. It then uses the past behavior of its users — their clicks, reads, views, comments, and so on — to curate a highly personalized news-feed tailored to each person's interests. The app's algorithms even rewrite headlines to optimize for user clicks. And the more those users click, the better Toutiao becomes at recommending precisely the content they want to see. It's a positive feedback loop that has created one of the most addictive content platforms on the internet, with users spending an average of seventy-four minutes per day in the app.

ROBOT REPORTS AND FAKE NEWS

Reaching beyond simple curation, Toutiao also uses machine learning to create and police its content. During the 2016 Summer Olympics in Rio de Janeiro, Toutiao worked with Peking University to create an AI "reporter" that wrote short articles summing up sports events within minutes of the final whistle. The writing wasn't exactly poetry, but the speed was incredible: the "reporter" produced short summaries within two seconds of some events' finish, and it "covered" over thirty events per day.

Algorithms are also being used to sniff out "fake news" on the platform, often in the form of bogus medical treatments. Originally, readers discovered and reported misleading stories—essentially, free labeling of that data. Toutiao then used that labeled data to train an algorithm that could identify fake news in the wild. Toutiao even trained a separate algorithm to *write* fake news stories. It then pitted those two algorithms against each other, competing to fool one another and improving both in the process.

This AI-driven approach to content is paying off. By late 2017, Toutiao was already valued at $20 billion and went on to raise a new round of funding that would value it at $30 billion, dwarfing the $1.7 billion valuation for BuzzFeed at the time. For 2018, Toutiao projected revenues between $4.5 and $7.6 billion. And the Chinese company is rapidly working to expand overseas. After trying and failing in 2016 to buy Reddit, the popular U.S. aggregation and discussion site, in 2017 Toutiao snapped up a France-based news aggregator and Musical.ly, a Chinese video lip-syncing app that's wildly popular with American teens.

Toutiao is just one company, but its success is indicative of China's strength in internet AI. With more than 700 million internet users all digesting content in the same language, China's internet juggernauts are reaping massive rewards from optimizing online services with AI. That has helped fuel the rapid rise of Tencent's market cap—surpassing Facebook in November 2017 and becoming the first Chinese company to top $500 billion—and has allowed Alibaba to hold its own with Amazon. Despite Baidu's strength in AI research, its mobile services lagged far behind Google. But that gap is more than made up for by upstarts like Toutiao, Chinese companies that are generating multibillion-dollar valuations by building their business foundation on internet AI. Massive profits will accrue to these internet companies as they become even better at holding our attention longer and harvesting our clicks.

Overall, Chinese and American companies are on about equal footing in internet AI, with around 50–50 odds of leadership based on current technology. I predict that in five years' time, Chinese technology companies will have a slight advantage (60–40) when it comes to leading the world in internet AI and reaping the richest rewards from its implementation. Remember, China alone has more internet users than the United States and all of Europe combined, and those users are empowered to make frictionless mobile payments to content creators, O2O platforms, and other users. That combination is generating creative internet AI applications and opportunities for monetization unmatched anywhere else in the world. Add China's tenacious and well-funded entrepreneurs into the mix, and China has a strong—but not yet decisive—edge over Silicon Valley.

But for all the economic value that the first AI wave generates, it remains largely bottled up in the high-tech sector and digital world. Bringing the optimization power of AI to bear on more traditional companies in the wider economy comes during the second wave: business AI.

SECOND WAVE: BUSINESS AI

First-wave AI leverages the fact that internet users are automatically labeling data as they browse. Business AI takes advantage of the fact that traditional companies have also been automatically labeling huge quantities of data for decades. For instance, insurance companies have been covering accidents and catching fraud, banks have been issuing loans and documenting repayment rates, and hospitals have been keeping records of diagnoses and survival rates. All of these actions generate labeled data points—a set of characteristics and a meaningful outcome—but until recently, most traditional businesses had a hard time exploiting that data for better results.

Business AI mines these databases for hidden correlations that often escape the naked eye and human brain. It draws on all the historic decisions and outcomes within an organization and uses labeled data to train an algorithm that can outperform even the most experienced human practitioners. That's because humans normally make predictions on the basis of *strong features,* a handful of data points that are highly correlated to a specific outcome, often in a clear cause-and-effect relationship. For example, in predicting the likelihood of someone contracting diabetes, a person's weight and body mass index are strong features. AI algorithms do indeed factor in these strong features, but they also look at thousands of other *weak features:* peripheral data points that might appear unrelated to the outcome but contain some predictive power when combined across tens of millions of examples. These subtle correlations are often impossible for any human to explain in terms of cause and effect: why do borrowers who take out loans on a Wednesday repay those loans faster? But algorithms that can combine thousands of those weak and strong features — often using complex mathematical relationships indecipherable to a human brain — will outperform even top-notch humans at many analytical business tasks.

Optimizations like this work well in industries with large amounts of structured data on meaningful business outcomes. In this case, "structured" refers to data that has been categorized, labeled, and made searchable. Prime examples of well-structured corporate data sets include historic stock prices, credit-card usage, and mortgage defaults.

THE BUSINESS OF BUSINESS AI

As early as 2004, companies like Palantir and IBM Watson offered big-data business consulting to companies and governments. But the widespread adoption of deep learning in 2013 turbocharged these capabilities and gave birth to new competitors, such as Element AI in Canada and 4th Paradigm in China.

These startups sell their services to traditional companies or organizations, offering to let their algorithms loose on existing databases in search of optimizations. They help these companies improve fraud detection, make smarter trades, and uncover inefficiencies in supply chains. Early instances of business AI have clustered heavily in the financial sector because it naturally lends itself to data analysis. The industry runs on well-structured information and has clear metrics that it seeks to optimize.

This is also why the United States has built a strong lead in early applications of business AI. Major American corporations already collect large amounts of data and store it in well-structured formats. They often use enterprise software for accounting, inventory, and customer relationship management. Once the data is in these formats, it's easy for companies like Palantir to come in and generate meaningful results by applying business AI to seek out cost savings and profit maximization.

This is not so in China. Chinese companies have never truly embraced enterprise software or standardized data storage, instead keeping their books according to their own idiosyncratic systems. Those systems are often not scalable and are difficult to integrate into existing software, making the cleaning and structuring of data a far more taxing process. Poor data also makes the results of AI optimizations less robust. As a matter of business culture, Chinese companies spend far less money on third-party consulting than their American counterparts. Many old-school Chinese businesses are still run more like personal fiefdoms than modern organizations, and outside expertise isn't considered something worth paying for.

FIRE YOUR BANKER

Both China's corporate data and its corporate culture make applying second-wave AI to its traditional companies a challenge. But in industries where business AI can leapfrog legacy systems, China is making serious strides. In these instances, China's relative backwardness in areas like financial services turns into a springboard to cutting-edge AI applications. One of the most promising of these is AI-powered micro-finance.

For example, when China leapfrogged credit cards to move right into mobile payments, it forgot one key piece of the consumer puzzle: credit itself. WeChat and Alipay let you draw directly from your bank account, but their core services don't give you the ability to spend a little bit beyond your means while you're waiting for the next paycheck.

Into this void stepped Smart Finance, an AI-powered app that relies exclusively on algorithms to make millions of small loans. Instead of asking borrowers to enter how much money they make, it simply requests access to some of the data on a potential borrower's phone. That data forms a kind of digital fingerprint, one with an astonishing ability to predict whether the borrower will pay back a loan of three hundred dollars.

Smart Finance's deep-learning algorithms don't just look to the obvious metrics, like how much money is in your WeChat Wallet. Instead, it derives predictive power from data points that would seem irrelevant to a human loan officer. For instance, it considers the speed at which you typed in your date of birth, how much battery power is left on your phone, and thousands of other parameters.

What does an applicant's phone battery have to do with creditworthiness? This is the kind of question that can't be answered in terms of simple cause and effect. But that's not a sign of the limitations of AI. It's a sign of the limitations of our own minds at recognizing correlations hidden within massive streams of data. By training its algorithms on millions of loans — many that got paid back and some that didn't — Smart Finance has discovered thousands of weak features that are correlated to creditworthiness, even if those correlations can't be explained in a simple way humans can understand. Those offbeat metrics constitute what Smart Finance founder Ke Jiao calls "a new standard of beauty" for lending, one to replace the crude metrics of income, zip code, and even credit score.

Growing mountains of data continue to refine these algorithms, allowing the company to scale up and extend credit to groups routinely ignored by China's traditional banking sector: young people and migrant workers. In late 2017, the company was making more than 2 million loans per month with default rates in the low single digits, a track record that makes traditional brick-and-mortar banks extremely jealous.

"THE ALGORITHM WILL SEE YOU NOW"

But business AI can be about more than dollars and cents. When applied to other information-driven public goods, it can mean a massive democratization of high-quality services to those who previously couldn't afford them. One of the most promising of these is medical diagnosis. Top researchers in the United States like Andrew Ng and Sebastian Thrun have demonstrated excellent algorithms that are on par with doctors at diagnosing specific illnesses based on images — pneumonia through chest x-rays and skin cancer through photos. But a broader business AI application for medicine will look to handle the entire diagnosis process for a wide variety of illnesses.

Right now, medical knowledge — and thus the power to deliver accurate diagnoses — is pretty much kept bottled up within a small number of very talented humans, people with imperfect memories and limited time to keep up with new advances in the field. Sure, a vast wealth of medical information is scattered across the internet but not in a way that is navigable by most people. First-rate medical diagnosis is still heavily rationed based on geography and, quite candidly, one's ability to pay.

This is especially stark in China, where well-trained doctors all cluster in the wealthiest cities. Travel outside of Beijing and Shanghai, and you're likely to see a dramatic drop in the medical knowledge of doctors treating your illness. The result? Patients from all around the country try to cram into the major hospitals, lining up for days and straining limited resources to the breaking point.

Second-wave AI promises to change all of this. Underneath the many social elements of visiting a doctor, the crux of diagnosis involves collecting data (symptoms, medical history, environmental factors) and predicting the phenomena correlated with them (an illness). This act of seeking out various correlations and making predictions is exactly what deep learning excels at. Given enough training data — in this case, precise medical records — an AI-powered diagnostic tool could turn any medical professional into a super-diagnostician, a doctor with experience in tens of millions of cases, an uncanny ability to spot hidden correlations, and a perfect memory to boot.

This is what RXThinking is attempting to build. Founded by a Chinese AI researcher with deep experience in Silicon Valley and at Baidu, the startup is training medical AI algorithms to become super-diagnosticians that can be dispatched to all corners of China. Instead of replacing doctors with algorithms, RXThinking's AI diagnosis app empowers them. It acts like a "navigation app" for the diagnosis process, drawing on all available knowledge to recommend the best route but still letting the doctors steer the car.

As the algorithm gains more information on each specific case, it progressively narrows the scope of possible illnesses and requests further clarifying information needed to complete the diagnosis. Once enough information has been entered to give the algorithm a high level of certainty, it makes a prediction for the cause of the symptoms, along with all other possible diagnoses and the percentage chance that they are the real culprit.

The app never overrides a doctor — who can always choose to deviate from the app's recommendations — but it draws on over 400 million existing medical records and continually scans the latest medical publications to make recommendations. It disseminates world-class medical knowledge equally throughout highly unequal societies, and lets all doctors and nurses focus on the human tasks that no machine can do: making patients feel cared for and consoling them when the diagnosis isn't bright.

JUDGING THE JUDGES

Similar principles are now being applied to China's legal system, another sprawling bureaucracy with highly uneven levels of expertise across regions. iFlyTek has taken the lead in applying AI to the courtroom, building tools and executing a Shanghai-based pilot program that uses data from past cases to advise judges on both evidence and sentencing. An evidence cross-reference system uses speech recognition and natural-language processing to compare all evidence presented — testimony, documents, and background material — and seek out contradictory fact patterns. It then alerts the judge to these disputes, allowing for further investigation and clarification by court officers.

Once a ruling is handed down, the judge can turn to yet another AI tool for advice on sentencing. The sentencing assistant starts with the fact pattern — defendant's criminal record, age, damages incurred, and so on — then its algorithms scan millions of court records for similar cases. It uses that body of knowledge to make recommendations for jail time or fines to be paid. Judges can also view similar cases as data points scattered across an X–Y graph, clicking on each dot for details on the fact pattern that led to the sentence. It's a process that builds consistency in a system with over 100,000 judges, and it can also rein in outliers whose sentencing patterns put them far outside the mainstream. One Chinese province is even using AI to rate and rank all prosecutors on their performance. Some American courts have implemented similar algorithms to advise on the "risk" level of prisoners up for parole, though the role and lack of transparency of these AI tools have already been challenged in higher courts.

As with RXThinking's "navigation system" for doctors, all of iFlyTek's judicial tools are just that: tools that aid a real human in making informed decisions. By empowering judges with data-driven recommendations, they can help balance the scales of justice and correct for the biases present in even well-trained judges. American legal scholars have illustrated vast disparities in U.S. sentencing based on the race of the victim and the defendant. And judicial biases can be far less malicious than racism: a study of Israeli judges found them far more severe in their decisions before lunch and more lenient in granting parole after having a good meal.

WHO LEADS?

So which country will lead in the broader category of business AI? Today, the United States enjoys a commanding lead (90–10) in this wave, but I believe in five years China will close that gap somewhat (70–30), and the Chinese government has a better shot at putting the power of business AI to good use. The United States has a clear advantage in the most immediate and profitable implementations of the technology: optimizations within banking, insurance, or any industry with lots of structured data that can be mined for better decision-making. Its companies have the raw material and corporate willpower to apply business AI to the problem of maximizing their bottom line.

There's no question that China will lag in the corporate world, but it may lead in public services and industries with the potential to leapfrog outdated systems. The country's immature financial system and imbalanced healthcare system give it strong incentives to rethink how services like consumer credit and medical care are distributed. Business AI will turn those weaknesses into strengths as it reimagines these industries from the ground up.

These applications of second-wave AI have immediate, real-world impacts, but the algorithms themselves are still trafficking purely in digital information mediated by humans. Third-wave AI changes all of this by giving AI two of humans' most valuable information-gathering tools: eyes and ears.

THIRD WAVE: PERCEPTION AI

Before AI, all machines were deaf and blind. Sure, you could take digital photos or make audio recordings, but these merely reproduced our audio and visual environments for humans to interpret—the machines themselves couldn't make sense of these reproductions. To a normal computer, a photograph is just a meaningless splattering of pixels it must store. To an iPhone, a song is just a series of zeros and ones that it must play for a human to enjoy.

This all changed with the advent of perception AI. Algorithms can now group the pixels from a photo or video into meaningful clusters and recognize objects in much the same way our brain does: golden retriever, traffic light, your brother Patrick, and so on. The same goes for audio data. Instead of merely storing audio files as collections of digital bits, algorithms can now both pick out words and often parse the meaning of full sentences.

Third-wave AI is all about extending and expanding this power throughout our lived environment, digitizing the world around us through the proliferation of sensors and smart devices. These devices are turning our physical world into digital data that can then be analyzed and optimized by deep-learning algorithms. Amazon Echo is digitizing the audio environment of people's homes. Alibaba's City Brain is digitizing urban traffic flows through cameras and object-recognition AI. Apple's iPhone X and Face++ cameras perform that same digitization for faces, using the perception data to safeguard your phone or digital wallet.

BLURRED LINES AND OUR "OMO" WORLD

As a result, perception AI is beginning to blur the lines separating the online and offline worlds. It does that by dramatically expanding the nodes through which we interact with the internet. Before perception AI, our interactions with the online world had to squeeze through two very narrow chokepoints: the keyboards on our computers or the screen on our smartphones. Those devices act as portals to the vast knowledge stored on the world wide web, but they are a very clunky way to input or retrieve information, especially when you're out shopping or driving in the real world.

As perception AI gets better at recognizing our faces, understanding our voices, and seeing the world around us, it will add millions of seamless points of contact between the online and offline worlds. Those nodes will be so pervasive that it no longer makes sense to think of oneself as "going online." When you order a full meal just by speaking a sentence from your couch, are you online or offline? When your refrigerator at home tells your shopping cart at the store that you're out of milk, are you moving through a physical world or a digital one?

I call these new blended environments OMO: *online-merge-of-fline*. OMO is the next step in an evolution that already took us from pure e-commerce deliveries to O2O (online-to-offline) services. Each of those steps has built new bridges between the online world and our physical one, but OMO constitutes the full integration of the two. It brings the convenience of the online world offline and the rich sensory reality of the offline world online. Over the coming years, perception AI will turn shopping malls, grocery stores, city streets, and our homes into OMO environments. In the process, it will produce some of the first applications of artificial intelligence that will feel truly futuristic to the average user.

Some of these are already here. One KFC restaurant in China recently teamed up with Alipay to pioneer a pay-with-your-face option at some stores. Customers place their own order at a digital terminal, and a quick facial scan connects their order to their Alipay account — no cash, cards, or cell phones required. The AI powering the machines even runs a quick "liveness algorithm" to ensure no one can use a photograph of someone else's face to pay for a meal.

Pay-with-your-face applications are fun, but they are just the tip of the OMO iceberg. To get a sense of where things are headed, let's take a quick trip just a few years into the future to see what a supermarket fully outfitted with perception AI devices might look like.

"WHERE EVERY SHOPPING CART KNOWS YOUR NAME"

"*Nihao,* Kai-Fu! Welcome back to Yonghui Superstore!"

It's always a nice feeling when your shopping cart greets you like an old friend. As I pull the cart back from the rack, visual sensors embedded in the handlebar have already completed a scan of my face and matched it to a rich, AI-driven profile of my habits, as a foodie, a shopper, and a husband to a fantastic cook of Chinese food. While I'm racking my brain for what groceries we'll need this week, a screen on the handlebar lights up.

"On the screen is a list of your typical weekly grocery purchase," the cart announces. And like that, our family's staple list of groceries appears on the screen: fresh eggplant, Sichuan pepper, Greek yogurt, skim milk, and so on.

My refrigerator and cabinets have already detected what items we're short on this week, and they automatically ordered the nonperishable staples — rice, soy sauce, cooking oil — for bulk delivery. That means grocery stores like Yonghui can tailor their selection around the items you'd want to pick out for yourself: fresh produce, unique wines, live seafood. It also allows the supermarkets to dramatically shrink their stores' footprint and place smaller stores within walking distance of most homes.

"Let me know if there's anything you'd like to add or subtract from the list," the cart chimes in. "Based on what's in your cart and your fridge at home, it looks like your diet will be short on fiber this week. Shall I add a bag of almonds or ingredients for a split-pea soup to correct that?"

"No split pea soup but have a large bag of almonds delivered to my house, thanks." I'm not sure an algorithm requires thanking, but I do it out of habit. Scanning the list, I make a couple of tweaks. My daughters are out of town so I can cut a few items, and I've already got some beef in my fridge so I decide to make my mother's recipe of beef noodles for my wife.

"Subtract the Greek yogurt and switch to whole milk from now on. Also, add the ingredients for beef noodles that I don't already have at home."

"No problem," it replies while adjusting my shopping list. The cart is speaking in Mandarin, but in the synthesized voice of my favorite actress, Jennifer Lawrence. It's a nice touch, and one of the reasons running errands doesn't feel like such a chore anymore.

The cart moves autonomously through the store, staying a few steps ahead of me while I pick out the ripest eggplants and the most fragrant Sichuan peppercorns, key to creating the numbing spice in the beef noodles. The cart then leads me to the back of the store where a precision-guided robot kneads and pulls fresh noodles for me. As I place them in the cart, depth-sensing cameras on the cart's rim recognize each item, and sensors lining the bottom weigh them as they go in.

The screen crosses things off as I go and displays the total cost. The precise location and presentation of every item has been optimized based on perception and purchase data gathered at the store: What displays do shoppers walk right by? Where do they stop and pick up items to inspect? And which of those do they finally purchase? That matrix of visual and business data gives AI-enabled supermarkets the same kind of rich understanding of consumer behavior that was previously reserved for online retailers.

Rounding the corner toward the wine aisle, a friendly young man in a concierge uniform approaches.

"Hi, Mr. Lee, how've you been?" he says. "We've just got in a shipment of some fantastic Napa wines. I understand that your wife's birthday is coming up, and we wanted to offer you a 10 percent discount on your first purchase of the 2014 Opus One. Your wife normally goes for Overture, and this is the premium offering from that same winery. It has some wonderful flavors, hints of coffee and even dark chocolate. Would you like a tasting?"

He knows my weakness for California wines, and I take him up on the offer. It's indeed fantastic.

"I love it," I say, handing the wineglass to the young man. "I'll take two bottles."

"Excellent choice — you can continue with your shopping, and I'll bring those bottles to you in just a moment. If you'd like to schedule regular deliveries to your home or need recommendations on what else to try, you can find those in the Yonghui app or with me here."

All the concierges are knowledgeable, friendly, and trained in the art of the upsell. It's far more socially engaged work than traditional supermarket jobs, with all employees ready to discuss recipes, farm-to-table sourcing, and how each product compares with what I've tried in the past.

The shopping trip goes on like this, with my cart leading me through our typical purchases, and concierges occasionally nudging me to splurge on items that algorithms predict I'll like. As a concierge is bagging my goods, my phone buzzes with this trip's receipt in my WeChat Wallet. When they're finished, the shopping cart guides itself back to its rack, and I stroll the two blocks home to my family.

Perception AI–powered shopping trips like this will capture one of the fundamental contradictions of the AI age before us: it will feel both completely ordinary and totally revolutionary. Much of our daily activity will still follow our everyday established patterns, but the digitization of the world will eliminate common points of friction and tailor services to each individual. They will bring the convenience and abundance of the online world into our offline reality. Just as important, by understanding and predicting the habits of each shopper, these stores will make major improvements in their supply chains, reducing food waste and increasing profitability.

And a supermarket like the one I've described isn't far off. The core technologies already exist, and it's largely a matter now of working out minor kinks in the software, integrating the back end of the supply chain, and building out the stores themselves.

AN OMO-POWERED EDUCATION

These kinds of immersive OMO scenarios go far beyond shopping. These same techniques — visual identification, speech recognition, creation of a detailed profile based on one's past behavior — can be used to create a highly tailored experience in education.

Present-day education systems are still largely run on the nineteenth-century "factory model" of education: all students are forced to learn at the same speed, in the same way, at the same place, and at the same time. Schools take an "assembly line" approach, passing children from grade to grade each year, largely irrespective of whether or not they absorbed what was taught. It's a model that once made sense given the severe limitations on teaching resources, namely, the time and attention of someone who can teach, monitor, and evaluate students.

But AI can help us lift those limitations. The perception, recognition, and recommendation abilities of AI can tailor the learning process to each student and also free up teachers for more one-on-one instruction time.

The AI-powered education experience takes place across four scenarios: in-class teaching, homework and drills, tests and grading, and customized tutoring. Performance and behavior in these four settings all feed into and build off of the bedrock of AI-powered education, the student profile. That profile contains a detailed accounting of everything that affects a student's learning process, such as what concepts they already grasp well, what they struggle with, how they react to different teaching methods, how attentive they are during class, how quickly they answer questions, and what incentives drive them. To see how this data is gathered and used to upgrade the education process, let's look at the four scenarios described above.

During in-class teaching, schools will employ a dual-teacher model that combines a remote broadcast lecture from a top educator and more personal attention by the in-class teacher. For the first half of class, a top-rated teacher delivers a lecture via a large-screen television at the front of the class. That teacher lectures simultaneously to around twenty classrooms and asks questions that students must answer via handheld clickers, giving the lecturer real-time feedback on whether students comprehend the concepts.

During the lecture, a video conference camera at the front of the room uses facial recognition and posture analysis to take attendance, check for student attentiveness, and assess the level of understanding based on gestures such as nodding, shaking one's head, and expressions of puzzlement. All of this data — answers to clicker questions, attentiveness, comprehension — goes directly into the student profile, filling in a real-time picture of what the students know and what they need extra help with.

But in-class learning is just a fraction of the whole AI-education picture. When students head home, the student profile combines with question-generating algorithms to create homework assignments precisely tailored to the students' abilities. While the whiz kids must complete higher-level problems that challenge them, the students who have yet to fully grasp the material are given more fundamental questions and perhaps extra drills.

At each step along the way, students' time and performance on different problems feed into their student profiles, adjusting the subsequent problems to reinforce understanding. In addition, for classes such as English (which is mandatory in Chinese public schools), AI-powered speech recognition can bring top-flight English instruction to the most remote regions. High-performance speech recognition algorithms can be trained to assess students' English pronunciation, helping them improve intonation and accent without the need for a native English speaker on site.

From a teacher's perspective, these same tools can be used to alleviate the burden of routine grading tasks, freeing up teachers to spend more time on the students themselves. Chinese companies have already used perception AI's visual recognition abilities to build scanners that can grade multiple-choice and fill-in-the-blank tests. Even in essays, standard errors such as spelling or grammar can be marked automatically, with predetermined deductions of points for certain mistakes. This AI-powered technology will save teachers' time in correcting the basics, letting them shift that time to communicating with students about higher-level writing concepts.

Finally, for students who are falling behind, the AI-powered student profile will notify parents of their child's situation, giving a clear and detailed explanation of what concepts the student is struggling with. The parents can use this information to enlist a remote tutor through services such as VIPKid, which connects American teachers with Chinese students for online English classes. Remote tutoring has been around for some time, but perception AI now allows these platforms to continuously gather data on student engagement through expression and sentiment analysis. That data continually feeds into a student's profile, helping the platforms filter for the kinds of teachers that keep students engaged.

Almost all of the tools described here already exist, and many are being implemented in different classrooms across China. Taken together, they constitute a new AI-powered paradigm for education, one that merges the online and offline worlds to create a learning experience tailored to the needs and abilities of each student. China appears poised to leapfrog the United States in education AI, in large part due to voracious demand from Chinese parents. Chinese parents of only children pour money into their education, a result of deeply entrenched Chinese values, intense competition for university spots, and a public education system of mixed quality. Those parents have already driven services like VIPKid to a valuation of over $3 billion in just a few years' time.

PUBLIC SPACES AND PRIVATE DATA

Creating and leveraging these OMO experiences requires vacuuming up oceans of data from the real world. Optimizing traffic flows via Alibaba's City Brain requires slurping up video feeds from around the city. Tailoring OMO retail experiences for each shopper requires identifying them via facial recognition. And accessing the power of the internet via voice commands requires technology that listens to our every word.

That type of data collection may rub many Americans the wrong way. They don't want Big Brother or corporate America to know too much about what they're up to. But people in China are more accepting of having their faces, voices, and shopping choices captured and digitized. This is another example of the broader Chinese willingness to trade some degree of privacy for convenience. That surveillance filters up from individual users to entire urban environments. Chinese cities already use a dense network of cameras and sensors to enforce traffic laws. That web of surveillance footage is now feeding directly into optimization algorithms for traffic management, policing, and emergency services.

It's up to each country to make its own decisions on how to balance personal privacy and public data. Europe has taken the strictest approach to data protection by introducing the General Data Protection Regulation, a law that sets a variety of restrictions on the collection and use of data within the European Union. The United States continues to grapple with implementing appropriate protections to user privacy, a tension illustrated by Facebook's Cambridge Analytica scandal and subsequent congressional hearings. China began implementing its own Cybersecurity Law in 2017, which included new punishments for the illegal collection or sale of user data.

There's no right answer to questions about what level of social surveillance is a worthwhile price for greater convenience and safety, or what level of anonymity we should be guaranteed at airports or subway stations. But in terms of immediate impact, China's relative openness with data collection in public places is giving it a massive head start on implementation of perception AI. It is accelerating the digitization of urban environments and opening the door to new OMO applications in retail, security, and transportation.

But pushing perception AI into these spheres requires more than just video cameras and digital data. Unlike internet and business AI, perception AI is a hardware-heavy enterprise. As we turn hospitals, cars, and kitchens into OMO environments, we will need a diverse array of sensor-enabled hardware devices to sync up the physical and digital worlds.

MADE IN SHENZHEN

Silicon Valley may be the world champion of software innovation, but Shenzhen (pronounced "shun-jun") wears that crown for hardware. In the last five years, this young manufacturing metropolis on China's southern coast has turned into the world's most vibrant ecosystem for building intelligent hardware. Creating an innovative app requires almost no real-world tools: all you need is a computer and a programmer with a clever idea. But building the hardware for perception AI—shopping carts with eyes and stereos with ears—demands a powerful and flexible manufacturing ecosystem, including sensor suppliers, injection-mold engineers, and small-batch electronics factories.

When most people think of Chinese factories, they envision

sweatshops with thousands of underpaid workers stitching together cheap shoes and teddy bears. These factories do still exist, but the Chinese manufacturing ecosystem has undergone a major technological upgrade. Today, the greatest advantage of manufacturing in China isn't the cheap labor — countries like Indonesia and Vietnam offer lower wages. Instead, it's the unparalleled flexibility of the supply chains and the armies of skilled industrial engineers who can make prototypes of new devices and build them at scale.

These are the secret ingredients powering Shenzhen, whose talented workers have transformed it from a dirt-cheap factory town to a go-to city for entrepreneurs who want to build new drones, robots, wearables, or intelligent machines. In Shenzhen, those entrepreneurs have direct access to thousands of factories and hundreds of thousands of engineers who help them iterate faster and produce goods cheaper than anywhere else.

At the city's dizzying electronics markets, they can choose from thousands of different variations of circuit boards, sensors, microphones, and miniature cameras. Once a prototype is assembled, the builders can go door to door at hundreds of factories to find one capable of producing their product in small batches or at large scale. That geographic density of parts suppliers and product manufacturers accelerates the innovation process. Hardware entrepreneurs say that a week spent working in Shenzhen is equivalent to a month in the United States.

As perception AI transforms our lived environment, the ease of experimentation and the production of smart devices gives Chinese startups an edge. Shenzhen is open to international hardware startups, but locals have a heavy home-court advantage. The many frictions of operating in a foreign country — language barrier, visa issues, tax complications, and distance from headquarters — can slow down American startups and raise the cost of their products. Massive multinationals like Apple have the resources to leverage Chinese manufacturing to the fullest, but for foreign startups small frictions can spell doom. Meanwhile, homegrown hardware startups in Shenzhen are like kids in a candy store, experimenting freely and building cheaply.

The Chinese hardware startup Xiaomi (pronounced "sheow-me") gives a glimpse of what a densely woven web of perception-AI devices could look like. Launched as a low-cost smartphone maker that took the country by storm, Xiaomi is now building a network of AI-empowered home devices that will turn our kitchens and living rooms into OMO environments.

Central to that system is the Mi AI speaker, a voice-command AI device similar to the Amazon Echo but at around half the price, thanks to the Chinese home-court manufacturing advantage. That advantage is then leveraged to build a range of smart, sensor-driven home devices: air purifiers, rice cookers, refrigerators, security cameras, washing machines, and autonomous vacuum cleaners. Xiaomi doesn't build all of these devices itself. Instead, it has invested in 220 companies and incubated 29 startups — many operating in Shenzhen — whose intelligent home products are hooked into the Xiaomi ecosystem. Together they are creating an affordable, intelligent home ecosystem, with WiFi-enabled products that find each other and make configuration easy. Xiaomi users can then simply control the entire ecosystem via voice command or directly on their phone.

It's a constellation of price, diversity, and capability that has created the world's largest network of intelligent home devices: 85 million by the end of 2017, far ahead of any comparable U.S. networks. It's also an ecosystem built on the Made-in-Shenzhen advantage. Low prices and China's massive market are turbocharging the data-gathering process for Xiaomi, fueling a virtuous cycle of stronger algorithms, smarter products, better user experience, more sales, and even more data. It's also an ecosystem that has produced four unicorn startups within Xiaomi's ecosystem alone and is driving Xiaomi toward an IPO predicted to value the company at around $100 billion.

As perception AI finds its way into more pieces of hardware, the entire home will feed into and operate off digitized real-world data. Your AI fridge will order more milk when it sees that you're running low. Your cappuccino machine will kick into gear at your voice command. The AI-equipped floors of your elderly parents will alert you immediately if they've tripped and fallen.

Third-wave AI products like these are on the verge of transforming our everyday environment, blurring lines between the digital and physical world until they disappear entirely. During this transformation, Chinese users' cultural nonchalance about data privacy and Shenzhen's strength in hardware manufacturing give it a clear edge in implementation. Today, China's edge is slight (60–40), but I predict that in five years' time, the above factors will give China a more than 80–20 chance of leading the United States and the rest of the world in the implementation of perception AI.

These third-wave AI innovations will create tremendous economic opportunities and also lay the foundation for the fourth and final wave, full autonomy.

FOURTH WAVE: AUTONOMOUS AI

Once machines can see and hear the world around them, they'll be ready to move through it safely and work in it productively. Autonomous AI represents the integration and culmination of the three preceding waves, fusing machines' ability to optimize from extremely complex data sets with their newfound sensory powers. Combining these superhuman powers yields machines that don't just understand the world around them — they shape it.

Self-driving cars may be on everyone's mind these days, but before we dive into autonomous vehicles, it's important to widen the lens and recognize just how deep and wide a footprint fourth-wave AI will have. Autonomous AI devices will revolutionize so much of our daily lives, including our malls, restaurants, cities, factories, and fire departments. As with the different waves of AI, this won't happen all at once. Early autonomous robotics applications will work only in highly structured environments where they can create immediate economic value. That means primarily factories, warehouses, and farms.

But aren't these places already highly automated? Hasn't heavy machinery already taken over many blue-collar line jobs? Yes, the developed world has largely replaced raw human muscle with high-powered machines. But while these machines are *automated,* they are not *autonomous.* While they can repeat an action, they can't make decisions or improvise according to changing conditions. Entirely blind to visual inputs, they must be controlled by a human or operate on a single, unchanging track. They can perform repetitive tasks, but they can't deal with any deviations or irregularities in the objects they manipulate. But by giving machines the power of sight, the sense of touch, and the ability to optimize from data, we can dramatically expand the number of tasks they can tackle.

STRAWBERRY FIELDS AND ROBOTIC BEETLES

Some of these applications are already at hand. Picking strawberries sounds like a straightforward task, but the ability to find, judge, and pluck fruits from plants proved impossible to automate before autonomous AI. Instead, tens of thousands of low-paid workers had to walk, hunched over, through strawberry fields all day, using their eyes and dexterous fingers to get the job done. It's grueling and tedious work, and many California farmers have watched fruit rot in their fields when they can't find people willing to take it on.

But the California-based startup Traptic has created a robot that can handle the task. The device is mounted on the back of a small tractor (or, in the future, an autonomous vehicle) and uses advanced vision algorithms to find the strawberries amid a sea of foliage. Those same algorithms check the color of the fruit to judge ripeness, and a machine arm delicately plucks them without any damage to the berry.

Amazon's warehouses give us an early glimpse of how transformative these technologies can be. Just five years ago, they looked like traditional warehouses: long aisles of sedentary shelves with humans walking or driving down the aisles to fetch inventory. Today, the humans stay put and the shelves come to them. Warehouses are covered with roving bands of autonomous beetle-like robots that scurry around with square-shaped towers of merchandise sitting on their backs. These beetles roam the factory floor, narrowly avoiding one another and bringing a handful of items to stationary humans when they need those goods. All the employees need to do is grab an item off that tower, scan it, and place it in a box. The humans stand in one place while the warehouse performs an elegantly choreographed autonomous ballet all around them.

All of these autonomous robots have one thing in common: they create direct economic value for their owners. As noted, autonomous AI will surface first in commercial settings because these robots create a tangible return on investment by doing the jobs of workers who are growing either more expensive or harder to find.

Domestic workers in the United States — cleaners, cooks, and caretakers — largely fit those criteria as well, but we're unlikely to see autonomous AI in the home any time soon. Counter to what sci-fi films have conditioned us to believe, human-like robots for the home remain out of reach. Seemingly simple tasks like cleaning a room or babysitting a child are far beyond AI's current capabilities, and our cluttered living environments constitute obstacle courses for clumsy robots.

SWARM INTELLIGENCE

But as autonomous technology becomes more agile and more intelligent, we will see some mind-bending and life-saving applications of the technology, particularly with drones. Swarms of autonomous drones will work together to paint the exterior of your house in just a few hours. Heat-resistant drone swarms will fight forest fires with hundreds of times the current efficiency of traditional fire crews. Other drones will perform search-and-rescue operations in the aftermath of hurricanes and earthquakes, bringing food and water to the stranded and teaming up with nearby drones to airlift people out.

Along these lines, China will almost certainly take the lead in autonomous drone technology. Shenzhen is home to DJI, the world's premier drone maker and what renowned tech journalist Chris Anderson called "the best company I have ever encountered." DJI is estimated to already own 50 percent of the North American drone market and even larger portions of the high-end segment. The company dedicates enormous resources to research and development, and is already deploying some autonomous drones for industrial and personal use. Swarm technologies are still in their infancy, but when hooked into Shenzhen's unmatched hardware ecosystem, the results will be awe-inspiring.

As these swarms transform our skies, autonomous cars will transform our roads. That revolution will also go far beyond transportation, disrupting urban environments, labor markets, and how we organize our days. Companies like Google have clearly demonstrated that self-driving cars will be far safer and more efficient than human drivers. Right now, dozens of startups, technology juggernauts, legacy carmakers, and electric vehicle makers are in an all-out sprint to be the first to truly commercialize the technology. Google, Baidu, Uber, Didi, Tesla, and many more are building teams, testing technologies, and gathering data en route to taking human drivers entirely out of the equation.

The leaders in that race — Google, through its self-driving spinoff Waymo, and Tesla — represent two different philosophies for autonomous deployment, two approaches with eerie echoes in the policies of the two AI superpowers.

THE GOOGLE APPROACH VERSUS THE TESLA APPROACH

Google was the first company to develop autonomous driving technology, but it has been relatively slow to deploy that technology at scale. Behind that caution is an underlying philosophy: build the perfect product and then make the jump straight to full autonomy once the system is far safer than human drivers. It's the approach of a perfectionist, one with a very low tolerance for risk to human lives or corporate reputation. It's also a sign of how large a lead Google has on the competition due to its multiyear head start on research. Tesla has taken a more incremental approach in an attempt to make up ground. Elon Musk's company has tacked on limited autonomous features to their cars as soon as they became available: autopilot for highways, autosteer for crash avoidance, and self-parking capabilities. It's an approach that accelerates speed of deployment while also accepting a certain level of risk.

The two approaches are powered by the same thing that powers

AI: data. Self-driving cars must be trained on millions, maybe billions, of miles of driving data so they can learn to identify objects and predict the movements of cars and pedestrians. That data draws from thousands of different vehicles on the road, and it all feeds into one central "brain," the core collection of algorithms that powers decision-making across the fleet. It means that when any autonomous car encounters a new situation, all the cars running on those algorithms learn from it.

Google has taken a slow-and-steady approach to gathering that data, driving around its own small fleet of vehicles equipped with very expensive sensing technologies. Tesla instead began installing cheaper equipment on its commercial vehicles, letting Tesla owners gather the data for them when they use certain autonomous features. The different approaches have led to a massive data gap between the two companies. By 2016, Google had taken six years to accumulate 1.5 million miles of real-world driving data. In just six months, Tesla had accumulated 47 million miles.

Google and Tesla are now inching toward one another in terms of approach. Google — perhaps feeling the heat from Tesla and other rivals — accelerated deployment of fully autonomous vehicles, piloting a program with taxi-like vehicles in the Phoenix metropolitan area. Meanwhile, Tesla appears to have pumped the brakes on its rapid rollout of fully autonomous vehicles, a deceleration that followed a May 2016 crash that killed a Tesla owner who was using autopilot.

But the fundamental difference in approach remains, and it presents a real tradeoff. Google is aiming for impeccable safety, but in the process it has delayed deployment of systems that could likely already save lives. Tesla takes a more techno-utilitarian approach, pushing their cars to market once they are an improvement over human drivers, hoping that the faster rates of data accumulation will train the systems earlier and save lives overall.

CHINA'S "TESLA" APPROACH

When managing a country of 1.39 billion people — one in which 260,000 people die in car accidents each year — the Chinese mentality is that you can't let the perfect be the enemy of the good. That is, rather than wait for flawless self-driving cars to arrive, Chinese leaders will likely look for ways to deploy more limited autonomous vehicles in controlled settings. That deployment will have the side effect of leading to more exponential growth in the accumulation of data and a corresponding advance in the power of the AI behind it.

Key to that incremental deployment will be the construction of new infrastructure specifically made to accommodate autonomous vehicles. In the United States, in contrast, we build self-driving cars to adapt to our existing roads because we assume the roads can't change. In China, there's a sense that everything can change — including current roads. Indeed, local officials are already modifying existing highways, reorganizing freight patterns, and building cities that will be tailor-made for driverless cars.

Highway regulators in the Chinese province of Zhejiang have already announced plans to build the country's first intelligent superhighway, infrastructure outfitted from the start for autonomous and electric vehicles. The plan calls for integrating sensors and wireless communication among the road, cars, and drivers to increase speeds by 20 to 30 percent and dramatically reduce fatalities. The superhighway will have photovoltaic solar panels built into the road surface, energy that feeds into charging stations for electric vehicles. In the long term, the goal is to be able to continuously charge electric vehicles while they drive. If successful, the project will accelerate deployment of autonomous and electric vehicles, leveraging the fact that long before autonomous AI can handle the chaos of urban driving, it can easily deal with highways — and gather more data in the process.

But Chinese officials aren't just adapting existing roads to autonomous vehicles. They're building entirely new cities around the technology. Sixty miles south of Beijing sits the Xiong'an New Area, a collection of sleepy villages where the central government has ordered the construction of a showcase city for technological progress and environmental sustainability. The city is projected to take in $583 billion worth of infrastructure spending and reach a population of 2.5 million, nearly as many people as Chicago. The idea of building a new Chicago from the ground up is fairly unthinkable in the United States, but in China it's just one piece of the government's urban planning toolkit.

Xiong'an is poised to be the world's first city built specifically to accommodate autonomous vehicles. Baidu has signed agreements with the local government to build an "AI City" with a focus on traffic management, autonomous vehicles, and environmental protection. Adaptations could include sensors in the cement, traffic lights equipped with computer vision, intersections that know the age of pedestrians crossing them, and dramatic reductions in space needed for parked cars. When everyone is hailing his or her own autonomous taxi, why not turn those parking lots into urban parks?

Taking things a step further, brand-new developments like Xiong'an could even route the traffic in their city centers underground, reserving the heart of town for pedestrians and bicyclists. It's a system that would be difficult, if not impossible, to implement in a world of human drivers prone to human errors that clog up tunnels. But by combining augmented roads, controlled lighting, and autonomous vehicles, an entire underground traffic grid could be running at the speed of highways while life aboveground moves at a more human pace.

There's no guarantee that all of these high-flying AI amenities will be rolled out smoothly — some of China's technologically themed developments have flopped, and some brand-new cities have struggled to attract residents. But the central government has placed a high priority on the project, and if successful, cities like Xiong'an will grow up together with autonomous AI. They will benefit from the efficiencies AI brings and will feed ever more data back into the algorithms. America's current infrastructure means that autonomous AI must adapt to and conquer the cities around it. In China, the government's proactive approach is to transform that conquest into co-evolution.

THE AUTONOMOUS BALANCE OF POWER

While all of this may sound exciting and innovative to the Chinese landscape, the hard truth is that no amount of government support can guarantee that China will lead in autonomous AI. When it comes to the core technology needed for self-driving cars, American companies remain two to three years ahead of China. In technology timelines, that's light-years of distance. Part of that stems from the relative importance of elite expertise in fourth-wave AI: safety issues and sheer complexity make autonomous vehicles a much tougher engineering nut to crack. It's a problem that requires a core team of world-class engineers rather than just a broad base of good ones. This tilts the playing field back toward the United States, where the best engineers from around the globe still cluster at companies like Google.

Silicon Valley companies also have a substantial head start on research and development, a product of the valley's proclivity for moonshot projects. Google began testing its self-driving cars as early as 2009, and many of its engineers went on to found early self-driving startups. China's boom in such startups really didn't begin until around 2016. Chinese giants like Baidu and autonomous-vehicle startups like Momenta, JingChi, and Pony.ai, however, are rapidly catching up in technology and data. Baidu's Apollo project — an open-source partnership and data-sharing arrangement among fifty autonomous-vehicle players, including chipmakers like Nvidia and automakers like Ford and Daimler — also presents an ambitious alternative to Waymo's closed, in-house approach. But even with that rapid catch-up by Chinese players, there's no question that as of this writing, the most experienced self-driving technologists still call America home.

Predicting which country takes the lead in autonomous AI largely comes down to one main question: will the primary bottleneck to full deployment be one of technology or policy? If the most intractable problems for deployment are merely technical ones, Google's Waymo has the best shot at solving them years ahead of the nearest competitor. But if new advances in fields like computer vision quickly disseminate throughout the industry — essentially, a rising technical tide lifting all boats — then Silicon Valley's head start on core technology may prove irrelevant. Many companies will become capable of building safe autonomous vehicles, and deployment will then become a matter of policy adaptation. In that universe, China's Tesla-esque policymaking will give its companies the edge.

At this point, we just don't yet know where that bottleneck will be, and fourth-wave AI remains anyone's game. While today the United States enjoys a commanding lead (90–10), in five years' time I give the United States and China even odds of leading the world in self-driving cars, with China having the edge in hardware-intensive applications such as autonomous drones. In the table below, I summarize my assessment of U.S. and Chinese capabilities across all four waves of AI, both in the present day and with my best estimate for how that balance will have evolved five years in the future.

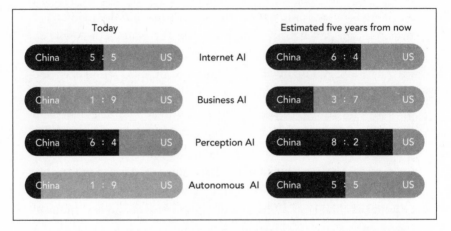

The balance of capabilities between the United States and China across the four waves of AI, currently and estimated for five years in the future

CONQUERING MARKETS AND ARMING INSURGENTS

What happens when you try to take these game-changing AI products global? Thus far, much of the work done in AI has been contained within the Chinese and U.S. markets, with companies largely avoiding direct competition on the home turf of the other nation. But despite the fact that the United States and China are the two largest economies in the world, the vast majority of AI's future users still live in other countries, many of them in the developing world. Any company that wants to be the Facebook or Google of the AI age needs a strategy for reaching those users and winning those markets.

Not surprisingly, Chinese and American tech companies are taking very different approaches to global markets: while America's global juggernauts seek to conquer these markets for themselves, China is instead arming the local startup insurgents.

In other words, Silicon Valley giants like Google, Facebook, and Uber want to directly introduce their products to these markets. They'll make limited efforts at localization but will largely stick to the traditional playbook. They will build one global product and push it out on billions of different users around the globe. It's an all-or-nothing approach with a huge potential upside if the conquest succeeds, but it also has a high chance of leaving empty-handed.

Chinese companies are instead steering clear of direct competition and investing in the scrappy local startups that Silicon Valley looks to wipe out. For example, in India and Southeast Asia, Alibaba and Tencent are pouring money and resources into homegrown startups that are fighting tooth and nail against juggernauts like Amazon. It's an approach rooted in the country's own native experience. People like Alibaba founder Jack Ma know how dangerous a ragtag bunch of insurgents can be when battling a monolithic foreign giant. So instead of seeking to both squash those startups and outcompete Silicon Valley, they're throwing their lot in with the locals.

RIDE-HAILING RUMBLE

There are already some precedents for the Chinese approach. Ever since Didi drove Uber out of China, it has invested in and partnered with local startups fighting to do the same thing in other countries: Lyft in the United States, Ola in India, Grab in Singapore, Taxify in Estonia, and Careem in the Middle East. After investing in Brazil's 99 Taxi in 2017, Didi outright acquired the company in early 2018. Together these startups have formed a global anti-Uber alliance, one that runs on Chinese money and benefits from Chinese know-how. After taking on Didi's investments, some of the startups have even rebuilt their apps in Didi's image, and others are planning to tap into Didi's strength in AI: optimizing driver matching, automatically adjudicating rider-driver disputes, and eventually rolling out autonomous vehicles.

We don't know the current depth of these technical exchanges, but they could serve as an alternate model of AI globalization: empower homegrown startups by marrying worldwide AI expertise to local data. It's a model built more on cooperation than conquest, and it may prove better suited to globalizing a technology that requires both top-quality engineers and ground-up data collection.

AI has a much higher localization quotient than earlier internet services. Self-driving cars in India need to learn the way pedestrians navigate the streets of Bangalore, and micro-lending apps in Brazil need to absorb the spending habits of millennials in Rio de Janeiro. Some algorithmic training can be transferred between different user bases, but there's no substitute for actual, real-world data.

Silicon Valley juggernauts do have some insight into the search and social habits in these countries. But building business, perception, and autonomous AI products will require companies to put real boots on the ground in each market. They will need to install hardware devices and localize AI services for the quirks of North African shopping malls and Indonesian hospitals. Projecting global power outward from Silicon Valley via computer code may not be the long-term answer.

Of course, no one knows the endgame for this global AI chess match. American companies could suddenly boost their localization efforts, leverage their existing products, and end up dominating all countries except China. Or a new generation of tenacious entrepreneurs in the developing world could use Chinese backing to create local empires impenetrable to Silicon Valley. If the latter scenario unfolds, China's tech giants wouldn't dominate the world, but they would play a role everywhere, improve their own algorithms using training data from many markets, and take home a substantial chunk of the profits generated.

LOOKING AHEAD

Scanning the AI horizon, we see waves of technology that will soon wash over the global economy and tilt the geopolitical landscape toward China. Traditional American companies are doing a good job of using deep learning to squeeze greater profits from their businesses,

and AI-driven companies like Google remain bastions of elite expertise. But when it comes to building new internet empires, changing
the way we diagnose illnesses, or reimagining how we shop, move,
and eat, China seems poised to seize global leadership. Chinese and
American internet companies have taken different approaches to
winning local markets, and as these AI services filter out to every
corner of the world, they may engage in proxy competition in countries like India, Indonesia, and parts of the Middle East and Africa.

This analysis sheds light on the emerging AI world order, but it
also showcases one of the blind spots in our AI discourse: the tendency to discuss it solely as a horse race. Who's ahead? What are the
odds for each player? Who's going to win?

This kind of competition matters, but if we dig deeper into the
coming changes, we find that far weightier questions lurk just below the surface. When the true power of artificial intelligence is
brought to bear, the real divide won't be between countries like the
United States and China. Instead, the most dangerous fault lines will
emerge within each country, and they will possess the power to tear
them apart from the inside.

6

★

UTOPIA, DYSTOPIA, AND
THE REAL AI CRISIS

All of the AI products and services outlined in the previous chapter are within reach based on current technologies. Bringing them to market requires no major new breakthroughs in AI research, just the nuts-and-bolts work of everyday implementation: gathering data, tweaking formulas, iterating algorithms in experiments and different combinations, prototyping products, and experimenting with business models.

But the age of implementation has done more than make these practical products possible. It has also set ablaze the popular imagination when it comes to AI. It has fed a belief that we're on the verge of achieving what some consider the Holy Grail of AI research, artificial general intelligence (AGI) — thinking machines with the ability to perform any intellectual task that a human can — and much more.

Some predict that with the dawn of AGI, machines that can improve themselves will trigger runaway growth in computer intelligence. Often called "the singularity," or artificial superintelligence, this future involves computers whose ability to understand and manipulate the world dwarfs our own, comparable to the intelligence gap between human beings and, say, insects. Such dizzying predictions have divided much of the intellectual community into two camps: utopians and dystopians.

The utopians see the dawn of AGI and subsequent singularity as the final frontier in human flourishing, an opportunity to expand our own consciousness and conquer mortality. Ray Kurzweil — the eccentric inventor, futurist, and guru-in-residence at Google — envisions a radical future in which humans and machines have fully merged. We will upload our minds to the cloud, he predicts, and constantly renew our bodies through intelligent nanobots released into our bloodstream. Kurzweil predicts that by 2029 we will have computers with intelligence comparable to that of humans (i.e., AGI), and that we will reach the singularity by 2045.

Other utopian thinkers see AGI as something that will enable us to rapidly decode the mysteries of the physical universe. DeepMind founder Demis Hassabis predicts that the creation of superintelligence will allow human civilization to solve intractable problems, producing inconceivably brilliant solutions to global warming and previously incurable diseases. With superintelligent computers that understand the universe on levels that humans cannot even conceive of, these machines become not just tools for lightening the burdens of humanity; they approach the omniscience and omnipotence of a god.

Not everyone, however, is so optimistic. Elon Musk has called superintelligence "the biggest risk we face as a civilization," comparing the creation of it to "summoning the demon." Intellectual celebrities such as the late cosmologist Stephen Hawking have joined Musk in the dystopian camp, many of them inspired by the work of Oxford philosopher Nick Bostrom, whose 2014 book *Superintelligence* captured the imagination of many futurists.

For the most part, members of the dystopian camp aren't worried about the AI takeover as imagined in films like the *Terminator* series, with human-like robots "turning evil" and hunting down people in a power-hungry conquest of humanity. Superintelligence would be the product of human creation, not natural evolution, and thus wouldn't have the same instincts for survival, reproduction, or domination that motivate humans or animals. Instead, it would likely just seek to achieve the goals given to it in the most efficient way possible.

The fear is that if human beings presented an obstacle to achieving one of those goals — reverse global warming, for example — a superintelligent agent could easily, even accidentally, wipe us off the face of the earth. For a computer program whose intellectual imagination so dwarfed our own, this wouldn't require anything as crude as gun-toting robots. Superintelligence's profound understanding of chemistry, physics, and nanotechnology would allow for far more ingenious ways to instantly accomplish its goals. Researchers refer to this as the "control problem" or "value alignment problem," and it's something that worries even AGI optimists.

Although timelines for these capabilities vary widely, Bostrom's book presents surveys of AI researchers, giving a median prediction of 2040 for the creation of AGI, with superintelligence likely to follow within three decades of that. But read on.

REALITY CHECK

When utopian and dystopian visions of the superintelligent future are discussed publicly, they inspire both awe and a sense of dread in audiences. Those all-consuming emotions then blur the lines in our mind separating these fantastical futures from our current age of AI implementation. The result is widespread popular confusion over where we truly stand today and where things are headed.

To be clear, *none* of the scenarios described above — the immortal digital minds or omnipotent superintelligences — are possible based on today's technologies; there remain no known algorithms for AGI or a clear engineering route to get there. The singularity is not something that can occur spontaneously, with autonomous vehicles running on deep learning suddenly "waking up" and realizing that they can band together to form a superintelligent network.

Getting to AGI would require a series of foundational scientific breakthroughs in artificial intelligence, a string of advances on the scale of, or greater than, deep learning. These breakthroughs would need to remove key constraints on the "narrow AI" programs that we run today and empower them with a wide array of new abilities: multidomain learning; domain-independent learning; natural-language understanding; commonsense reasoning, planning, and learning from a small number of examples. Taking the next step to emotionally intelligent robots may require self-awareness, humor, love, empathy, and appreciation for beauty. These are the key hurdles that separate what AI does today — spotting correlations in data and making predictions — and artificial general intelligence. Any one of these new abilities may require multiple huge breakthroughs; AGI implies solving all of them.

The mistake of many AGI forecasts is to simply take the rapid rate of advance from the past decade and extrapolate it outward or launch it exponentially upward in an unstoppable snowballing of computer intelligence. Deep learning represents a major leveling up in machine learning, a movement onto a new plateau with a variety of real-world uses: the age of implementation. But there is no proof that this upward change represents the beginning of exponential growth that will inevitably race toward AGI, and then superintelligence, at an ever-increasing pace.

Science is difficult, and fundamental scientific breakthroughs are even harder. Discoveries like deep learning that truly raise the bar for machine intelligence are rare and often separated by decades, if not longer. Implementations and improvements on these breakthroughs abound, and researchers at places like DeepMind have demonstrated powerful new approaches to things like reinforcement learning. But in the twelve years since Geoffrey Hinton and his colleagues' landmark paper on deep learning, I haven't seen anything that represents a similar sea change in machine intelligence. Yes, the AI scientists surveyed by Bostrom predicted a median date of 2040 for AGI, but I believe scientists tend to overestimate when an academic demonstration will become a real-world product. To wit, in the late 1980s, I was the world's leading researcher on AI speech recognition, and I joined Apple because I believed the technology would go mainstream within five years. It turned out that I was off by twenty years.

I cannot guarantee that scientists definitely will not make the breakthroughs that would bring about AGI and then superintelligence. In fact, I believe we should expect continual improvements to the existing state of the art. But I believe we are still many decades, if not centuries, away from the real thing. There is also a real possibility that AGI is something humans will never achieve. Artificial general intelligence would be a major turning point in the relationship between humans and machines — what many predict would be the most significant single event in the history of the human race. It's a milestone that I believe we should not cross unless we have first definitively solved all problems of control and safety. But given the relatively slow rate of progress on fundamental scientific breakthroughs, I and other AI experts, among them Andrew Ng and Rodney Brooks, believe AGI remains farther away than often imagined.

Does that mean I see nothing but steady material progress and glorious human flourishing in our AI future? Not at all. Instead, I believe that civilization will soon face a different kind of AI-induced crisis. This crisis will lack the apocalyptic drama of a Hollywood blockbuster, but it will disrupt our economic and political systems all the same, and even cut to the core of what it means to be human in the twenty-first century.

In short, this is the coming crisis of jobs and inequality. Our present AI capabilities can't create a superintelligence that destroys our civilization. But my fear is that we humans may prove more than up to that task ourselves.

FOLDING BEIJING: SCIENCE-FICTION VISIONS AND AI ECONOMICS

When the clock strikes 6 a.m., the city devours itself. Densely packed buildings of concrete and steel bend at the hip and twist at their spines. External balconies and awnings are turned inward, creating smooth and tightly sealed exteriors. Skyscrapers break down into component parts, shuffling and consolidating into Rubik's Cubes of industrial proportions. Inside those blocks are the residents of Beijing's Third Space, the economic underclass that toils during the night hours and sleeps during the day. As the cityscape folds in on itself, a patchwork of squares on the earth's surface begin their 180-degree rotation, flipping over to tuck these consolidated structures underground.

When the other side of these squares turn skyward, they reveal a separate city. The first rays of dawn creep over the horizon as this new city emerges from its crouch. Tree-lined streets, vast public parks, and beautiful single-family homes begin to unfold, spreading outward until they have covered the surface entirely. The residents of First Space stir from their slumber, stretching their limbs and looking out on a world all their own.

These are visions of Hao Jingfang, a Chinese science-fiction writer and economics researcher. Hao's novelette "Folding Beijing" won the prestigious Hugo Award in 2016 for its arresting depiction of a city in which economic classes are separated into different worlds.

In a futuristic Beijing, the city is divided into three economic castes that split time on the city's surface. Five million residents of the elite First Space enjoy a twenty-four-hour cycle beginning at 6 a.m., a full day and night in a clean, hypermodern, uncluttered city. When First Space folds up and flips over, the 20 million residents of Second Space get sixteen hours to work across a somewhat less glamorous cityscape. Finally, the denizens of Third Space — 50 million sanitation workers, food vendors, and menial laborers — emerge for an eight-hour shift from 10 p.m. to 6 a.m., toiling in the dark among the skyscrapers and trash pits.

The trash-sorting jobs that are a pillar of the Third Space could be entirely automated but are instead done manually to provide employment for the unfortunate denizens condemned to life there. Travel between the different spaces is forbidden, creating a society in which the privileged residents of First Space can live free of worry that the unwashed masses will contaminate their techno-utopia.

THE REAL AI CRISIS

This dystopian story is a work of science fiction but one rooted in real fears about economic stratification and unemployment in our automated future. Hao holds a Ph.D. in economics and management from prestigious Tsinghua University. For her day job, she conducts economics research at a think tank reporting to the Chinese central government, including investigating the impact of AI on jobs in China.

It's a subject that deeply worries many economists, technologists, and futurists, myself included. I believe that as the four waves of AI spread across the global economy, they have the potential to wrench open ever greater economic divides between the haves and have-nots, leading to widespread technological unemployment. As Hao's story so vividly illustrates, these chasms in wealth and class can morph into something much deeper: economic divisions that tear at the fabric of our society and challenge our sense of human dignity and purpose.

Massive productivity gains will come from the automation of profit-generating tasks, but they will also eliminate jobs for huge numbers of workers. These layoffs won't discriminate by the color of one's collar, hitting highly educated white-collar workers just as hard as many manual laborers. A college degree — even a highly specialized professional degree — is no guarantee of job security when competing against machines that can spot patterns and make decisions on levels the human brain simply can't fathom.

Beyond direct job losses, artificial intelligence will exacerbate global economic inequality. By giving robots the power of sight and the ability to move autonomously, AI will revolutionize manufacturing, putting third-world sweatshops stocked with armies of low-wage workers out of business. In doing so, it will cut away the bottom rungs on the ladder of economic development. It will deprive poor countries of the opportunity to kick-start economic growth through low-cost exports, the one proven route that has lifted countries like South Korea, China, and Singapore out of poverty. The large populations of young workers that once comprised the greatest advantage of poor countries will turn into a net liability, and a potentially destabilizing one. With no way to begin the development process, poor countries will stagnate while the AI superpowers take off.

But even within those rich and technologically advanced countries, AI will further cleave open the divide between the haves and the have-nots. The positive-feedback loop generated by increasing amounts of data means that AI-driven industries naturally tend toward monopoly, simultaneously driving down prices and eliminating competition among firms. While small businesses will ultimately be forced to close their doors, the industry juggernauts of the AI age will see profits soar to previously unimaginable levels. This concentration of economic power in the hands of a few will rub salt in the open wounds of social inequality.

In most developed countries, economic inequality and class-based resentment rank among the most dangerous and potentially explosive problems. The past few years have shown us how a cauldron of long-simmering inequality can boil over into radical political upheaval. I believe that, if left unchecked, AI will throw gasoline on the socioeconomic fires.

Lurking beneath this social and economic turmoil will be a psychological struggle, one that won't make the headlines but that could make all the difference. As more and more people see themselves displaced by machines, they will be forced to answer a far deeper question: in an age of intelligent machines, what does it mean to be human?

THE TECHNO-OPTIMISTS AND
THE "LUDDITE FALLACY"

Like the utopian and dystopian forecasts for AGI, this prediction of a jobs and inequality crisis is not without controversy. A large contingent of economists and techno-optimists believe that fears about technology-induced job losses are fundamentally unfounded.

Members of this camp dismiss dire predictions of unemployment as the product of a "Luddite fallacy." The term is derived from the Luddites, a group of nineteenth-century British weavers who smashed the new industrial textile looms that they blamed for destroying their livelihoods. Despite the best efforts and protests of the Luddites, industrialization plowed full steam ahead, and both the number of jobs and quality of life in England rose steadily for much of the next two centuries. The Luddites may have failed in their bid to protect their craft from automation — and many of those directly impacted by automation did in fact suffer stagnant wages for some time — but their children and grandchildren were ultimately far better off for the change.

This, the techno-optimists assert, is the real story of technological change and economic development. Technology improves human productivity and lowers the price of goods and services. Those lower prices mean consumers have greater spending power, and they either buy more of the original goods or spend that money on something else. Both of these outcomes increase the demand for labor and thus jobs. Yes, shifts in technology might lead to some short-term displacement. But just as millions of farmers became factory workers, those laid-off factory workers can become yoga teachers and software programmers. Over the long term, technological progress never truly leads to an actual reduction in jobs or rise in unemployment.

It's a simple and elegant explanation of the ever-increasing material wealth and relatively stable job markets in the industrialized world. It also serves as a lucid rebuttal to a series of "boy who cried wolf" moments around technological unemployment. Ever since the Industrial Revolution, people have feared that everything from weaving looms to tractors to ATMs will lead to massive job losses. But each time, increasing productivity has paired with the magic of the market to smooth things out.

Economists who look to history — and the corporate juggernauts who will profit tremendously from AI — use these examples from the past to dismiss claims of AI-induced unemployment in the future. They point to millions of inventions — the cotton gin, lightbulbs, cars, video cameras, and cell phones — none of which led to widespread unemployment. Artificial intelligence, they say, will be no different. It will greatly increase productivity and promote healthy growth in jobs and human welfare. So what is there to worry about?

THE END OF BLIND OPTIMISM

If we think of all inventions as data points and weight them equally, the techno-optimists have a compelling and data-driven argument. But not all inventions are created equal. Some of them change how we perform a single task (typewriters), some of them eliminate the need for one kind of labor (calculators), and some of them disrupt a whole industry (the cotton gin).

And then there are technological changes on an entirely different scale. The ramifications of these breakthroughs will cut across dozens of industries, with the potential to fundamentally alter economic processes and even social organization. These are what economists call general purpose technologies, or GPTs. In their landmark book *The Second Machine Age,* MIT professors Erik Brynjolfsson and Andrew McAfee described GPTs as the technologies that "really matter," the ones that "interrupt and accelerate the normal march of economic progress."

Looking only at GPTs dramatically shrinks the number of data points available for evaluating technological change and job losses. Economic historians have many quibbles over exactly which innovations of the modern era should qualify (railroads? the internal combustion engine?), but surveys of the literature reveal three technologies that receive broad support: the steam engine, electricity, and information and communication technology (such as computers and the internet). These have been the game changers, the disruptive technologies that extended their reach into many corners of the economy and radically altered how we live and work.

These three GPTs have been rare enough to warrant evaluation on their own, not simply to be lumped in with millions of more narrow innovations like the ballpoint pen or automatic transmission. And while it's true that the long-term historical trend has been toward more jobs and greater prosperity, when looking at GPTs alone, three data points are not enough to extract an ironclad principle. Instead, we should look to the historical record to see how each of these groundbreaking innovations has affected jobs and wages.

The steam engine and electrification were crucial pieces of the first and second Industrial Revolutions (1760–1830 and 1870–1914, respectively). Both of these GPTs facilitated the creation of the modern factory system, bringing immense power and abundant light to the buildings that were upending traditional modes of production. Broadly speaking, this change in the mode of production was one of *deskilling*. These factories took tasks that once required high-skilled workers (for example, handcrafting textiles) and broke the work down into far simpler tasks that could be done by low-skilled workers (operating a steam-driven power loom). In the process, these technologies greatly increased the amount of these goods produced and drove down prices.

In terms of employment, early GPTs enabled process innovations like the assembly line, which gave thousands — and eventually hundreds of millions — of former farmers a productive role in the new industrial economy. Yes, they displaced a relatively small number of skilled craftspeople (some of whom would become Luddites), but they empowered much larger numbers of low-skilled workers to take on repetitive, machine-enabled jobs that increased their productivity. Both the economic pie and overall standards of living grew.

But what about the most recent GPT, information and communication technologies (ICT)? So far, its impact on labor markets and wealth inequality have been far more ambiguous. As Brynjolfsson and McAfee point out in *The Second Machine Age,* over the past thirty years, the United States has seen steady growth in worker productivity but stagnant growth in median income and employment. Brynjolfsson and McAfee call this "the great decoupling." After decades when productivity, wages, and jobs rose in almost lockstep fashion, that once tightly woven thread has begun to fray. While productivity has continued to shoot upward, wages and jobs have flatlined or fallen.

This has lead to growing economic stratification in developed countries like the United States, with the economic gains of ICT increasingly accruing to the top 1 percent. That elite group in the United States has roughly doubled its share of national income between 1980 and 2016. By 2017, the top 1 percent of Americans possessed almost twice as much wealth as the bottom 90 percent combined. While the most recent GPT proliferated across the economy, real wages for the median of Americans have remained flat for over thirty years, and they've actually fallen for the poorest Americans.

One reason why ICT may differ from the steam engine and electrification is because of its "skill bias." While the two other GPTs ramped up productivity by *deskilling* the production of goods, ICT is instead often — though not always — *skill biased* in favor of high-skilled workers. Digital communications tools allow top performers to efficiently manage much larger organizations and reach much larger audiences. By breaking down the barriers to disseminating information, ICT empowers the world's top knowledge workers and undercuts the economic role of many in the middle.

Debates over how large a role ICT has played in job and wage stagnation in the United States are complex. Globalization, the decline of labor unions, and outsourcing are all factors here, providing economists with fodder for endless academic arguments. But one thing is increasingly clear: there is no guarantee that GPTs that increase our productivity will also lead to more jobs or higher wages for workers.

Techno-optimists can continue to dismiss these concerns as the same old Luddite fallacy, but they are now arguing against some of the brightest economic minds of today. Lawrence Summers has served as the chief economist of the World Bank, as the treasury secretary under President Bill Clinton, and as the director of President Barack Obama's National Economic Council. In recent years, he has been warning against the no-questions-asked optimism around technological change and employment.

"The answer is surely not to try to stop technical change," Summers told the *New York Times* in 2014, "but the answer is not to just suppose that everything's going to be O.K. because the magic of the market will assure that's true."

Erik Brynjolfsson has issued similar warnings about the growing disconnect between the creation of wealth and jobs, calling it "the biggest challenge of our society for the next decade."

AI: PUTTING THE G IN GPT

What does all this have to do with AI? I am confident that AI will soon enter the elite club of universally recognized GPTs, spurring a revolution in economic production and even social organization. The AI revolution will be on the scale of the Industrial Revolution, but probably larger and definitely faster. Consulting firm PwC predicts that AI will add $15.7 trillion to the global economy by 2030. If that prediction holds up, it will be an amount larger than the entire GDP of China today and equal to approximately 80 percent of the GDP of the United States in 2017. Seventy percent of those gains are predicted to accrue in the United States and China.

These disruptions will be more broad-based than prior economic revolutions. Steam power fundamentally altered the nature of manual labor, and ICT did the same for certain kinds of cognitive labor. AI will cut across both. It will perform many kinds of physical and intellectual tasks with a speed and power that far outstrip any human, dramatically increasing productivity in everything from transportation to manufacturing to medicine.

Unlike the GPTs of the first and second Industrial Revolutions, AI will not facilitate the deskilling of economic production. It won't take advanced tasks done by a small number of people and break them down further for a larger number of low-skill workers to do. Instead, it will simply take over the execution of tasks that meet two criteria: they can be optimized using data, and they do not require social interaction. (I will be going into greater detail about exactly which jobs AI can and cannot replace.)

Yes, there will be some new jobs created along the way — robot repairing and AI data scientists, for example. But the main thrust of AI's employment impact is not one of job creation through deskilling but of job replacement through increasingly intelligent machines. Displaced workers can theoretically transition into other industries that are more difficult to automate, but this is itself a highly disruptive process that will take a long time.

HARDWARE, BETTER, FASTER, STRONGER

And time is one thing that the AI revolution is not inclined to grant us. The transition to an AI-driven economy will be far faster than any of the prior GPT-induced transformations, leaving workers and organizations in a mad scramble to adjust. Whereas the Industrial Revolution took place across several generations, the AI revolution will have a major impact within one generation. That's because AI adoption will be accelerated by three catalysts that didn't exist during the introduction of steam power and electricity.

First, many productivity-increasing AI products are just digital algorithms: infinitely replicable and instantly distributable around the world. This makes for a stark contrast to the hardware-intensive revolutions of steam power, electricity, and even large parts of ICT. For these transitions to gain traction, physical products had to be invented, prototyped, built, sold, and shipped to end users. Each time a marginal improvement was made to one of these pieces of hardware, it required that the earlier process be repeated, with the attendant costs and social frictions that slowed down adoption of each new tweak. All of these frictions slowed down development of new technologies and extended the time until a product was cost-effective for businesses to adopt.

In contrast, the AI revolution is largely free of these limitations. Digital algorithms can be distributed at virtually no cost, and once distributed, they can be updated and improved for free. These algorithms — not advanced robotics — will roll out quickly and take a large chunk out of white-collar jobs. Much of today's white-collar workforce is paid to take in and process information, and then make a decision or recommendation based on that information — which is precisely what AI algorithms do best. In industries with a minimal social component, that human-for-machine replacement can be made rapidly and done en masse, without any need to deal with the messy details of manufacturing, shipping, installation, and on-site repairs. While the hardware of AI-powered robots or self-driving cars will bear some of these legacy costs, the underlying software does not, allowing for the sale of machines that actually get better over time. Lowering these barriers to distribution and improvement will rapidly accelerate AI adoption.

The second catalyst is one that many in the technology world today take for granted: the creation of the venture-capital industry. VC funding — early investments in high-risk, high-potential companies — barely existed before the 1970s. That meant the inventors and innovators during the first two Industrial Revolutions had to rely on a thin patchwork of financing mechanisms to get their products off the ground, usually via personal wealth, family members, rich patrons, or bank loans. None of these have incentive structures built for the high-risk, high-reward game of funding transformative innovation. That dearth of innovation financing meant many good ideas likely never got off the ground, and successful implementation of the GPTs scaled far more slowly.

Today, VC funding is a well-oiled machine dedicated to the creation and commercialization of new technology. In 2017, global venture funding set a new record with $148 billion invested, egged on by the creation of Softbank's $100 billion "vision fund," which will be disbursed in the coming years. That same year, global VC funding for AI startups leaped to $15.2 billion, a 141 percent increase over

2016. That money relentlessly seeks out ways to wring every dollar of productivity out of a GPT like artificial intelligence, with a particular fondness for moonshot ideas that could disrupt and recreate an entire industry. Over the coming decade, voracious VCs will drive the rapid application of the technology and the iteration of business models, leaving no stone unturned in exploring everything that AI can do.

Finally, the third catalyst is one that's equally obvious and yet often overlooked: China. Artificial intelligence will be the first GPT of the modern era in which China stands shoulder to shoulder with the West in both advancing and applying the technology. During the eras of industrialization, electrification, and computerization, China lagged so far behind that its people could contribute little, if anything, to the field. It's only in the past five years that China has caught up enough in internet technologies to feed ideas and talent back into the global ecosystem, a trend that has dramatically accelerated innovation in the mobile internet.

With artificial intelligence, China's progress allows for the research talent and creative capacity of nearly one-fifth of humanity to contribute to the task of distributing and utilizing artificial intelligence. Combine this with the country's gladiatorial entrepreneurs, unique internet ecosystem, and proactive government push, and China's entrance to the field of AI constitutes a major accelerant to AI that was absent for previous GPTs.

Reviewing the preceding arguments, I believe we can confidently state a few things. First, during the industrial era, new technology has been associated with long-term job creation and wage growth. Second, despite this general trend toward economic improvement, GPTs are rare and substantial enough that each one's impact on jobs should be evaluated independently. Third, of the three widely recognized GPTs of the modern era, the skill biases of steam power and electrification boosted both productivity and employment. ICT has lifted the former but not necessarily the latter, contributing to falling wages for many workers in the developed world and greater inequality. Finally, AI will be a GPT, one whose skill biases and speed of adoption — catalyzed by digital dissemination, VC funding, and

China — suggest it will lead to negative impacts on employment and income distribution.

If the above arguments hold true, the next questions are clear: What jobs are really at risk? And how bad will it be?

WHAT AI CAN AND CAN'T DO: THE RISK-OF-REPLACEMENT GRAPHS

When it comes to job replacement, AI's biases don't fit the traditional one-dimensional metric of low-skill versus high-skill labor. Instead, AI creates a mixed bag of winners and losers depending on the particular content of job tasks performed. While AI has far surpassed humans at narrow tasks that can be optimized based on data, it remains stubbornly unable to interact naturally with people or imitate the dexterity of our fingers and limbs. It also cannot engage in cross-domain thinking on creative tasks or ones requiring complex strategy, jobs whose inputs and outcomes aren't easily quantified. What this means for job replacement can be expressed simply through two X–Y graphs, one for physical labor and one for cognitive labor.

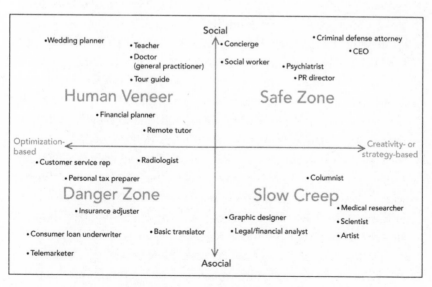

Risk of Replacement: Cognitive Labor

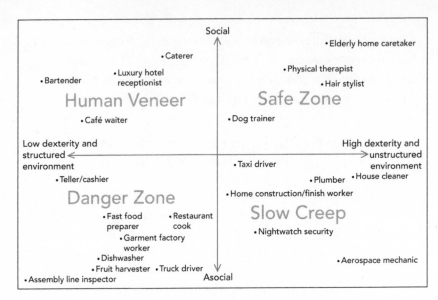

Risk of Replacement: Physical Labor

For physical labor, the X-axis extends from "low dexterity and structured environment" on the left side, to "high dexterity and unstructured environment" on the right side. The Y-axis moves from "asocial" at the bottom to "highly social" at the top. The cognitive labor chart shares the same Y-axis (asocial to highly social) but uses a different X-axis: "optimization-based" on the left, to "creativity- or strategy-based" on the right. Cognitive tasks are categorized as "optimization-based" if their core tasks involve maximizing quantifiable variables that can be captured in data (for example, setting an optimal insurance rate or maximizing a tax refund).

These axes divide both charts into four quadrants: the bottom-left quadrant is the "Danger Zone," the top-right is the "Safe Zone," the top-left is the "Human Veneer," and the bottom right is the "Slow Creep." Jobs whose tasks primarily fall in the "Danger Zone" (dishwasher, entry-level translators) are at a high risk of replacement in the coming years. Those in the "Safe Zone" (psychiatrist, home-care nurse, etc.) are likely out of reach of automation for the foreseeable future. The "Human Veneer" and "Slow Creep" quadrants are less clear-cut: while not fully replaceable right now, reorganization of work tasks or steady advances in technology could lead to widespread job reductions in these quadrants. As we will see, occupations often involve many different activities outside of the "core tasks" that we have used to place them in a given quadrant. This task-diversity will complicate the automation of many professions, but for now we can use these axes and quadrants as general guidance for thinking about what occupations are at risk.

For the "Human Veneer" quadrant, much of the computational or physical work can already be done by machines, but the key social interactive element makes them difficult to automate en masse. The name of the quadrant derives from the most likely route to automation: while the behind-the-scenes optimization work is overtaken by machines, human workers will act as the social interface for customers, leading to a symbiotic relationship between human and machine. Jobs in this category could include bartender, schoolteacher, and even medical caregiver. How quickly and what percentage of these jobs disappear depends on how flexible companies are in restructuring the tasks done by their employees, and how open customers are to interacting with computers.

The "Slow Creep" category (plumber, construction worker, entry-level graphic designer) doesn't rely on human beings' social skills but instead on manual dexterity, creativity, or ability to adapt to unstructured environments. These remain substantial hurdles for AI, but ones that the technology will slowly chip away at in the coming years. The pace of job elimination in this quadrant depends less on process innovation at companies and more on the actual expansion in AI capabilities. But at the far right end of the "Slow Creep" are good opportunities for the creative professionals (such as scientists and aerospace engineers) to use AI tools to accelerate their progress.

These graphs give us a basic heuristic for understanding what *kinds* of jobs are at risk, but what does this mean for *total employment* on an economy-wide level? For that, we must look to the economists.

WHAT THE STUDIES SAY

Predicting the scale of AI-induced job losses has become a cottage industry for economists and consulting firms the world over. Depending on which model one uses, estimates range from terrifying to totally not a problem. Here I give a brief overview of the literature and the methods, highlighting the studies that have shaped the debate. Few good studies have been done for the Chinese market, so I largely stick to studies estimating automation potential in the United States and then extrapolate those results to China.

A pair of researchers at Oxford University kicked things off in 2013 with a paper making a dire prediction: 47 percent of U.S. jobs could be automated within the next decade or two. The paper's authors, Carl Benedikt Frey and Michael A. Osborne, began by asking machine-learning experts to evaluate the likelihood that seventy occupations could be automated in the coming years. Combining that data with a list of the main "engineering bottlenecks" in machine learning (similar to the characteristics denoting the "Safe Zone" in the graphs on pages 155 and 156), Frey and Osborne used a probability model to project how susceptible an additional 632 occupations are to automation.

The result — that nearly half of U.S. jobs were at "high risk" in the coming decades — caused quite a stir. Frey and Osborn were careful to note the many caveats to their conclusion. Most importantly, it was an estimate of what jobs it would be *technically possible* to do with machines, not actual job losses or resulting unemployment levels. But the ensuing flurry of press coverage largely glossed over these important details, instead warning readers that half of all workers would soon be out of a job.

Other economists struck back. In 2016, a trio of researchers at the Organization for Economic Cooperation and Development (OECD) used an alternate model to produce an estimate that seemed to directly contradict the Oxford study: just 9 percent of jobs in the United States were at high risk of automation.

Why the huge gap? The OECD researchers took issue with Osborne and Frey's "occupation-based" approach. While the Oxford researchers asked machine-learning experts to judge the automatability of an occupation, the OECD team pointed out that it's not entire occupations that will be automated but rather specific tasks within those occupations. The OECD team argued that this focus on occupations overlooks the many different tasks an employee performs that an algorithm cannot: working with colleagues in groups, dealing with customers face-to-face, and so on.

The OECD team instead proposed a *task-based* approach, breaking down each job into its many component activities and looking at how many of those could be automated. In this model, a tax preparer is not merely categorized as one occupation but rather as a series of tasks that are automatable (reviewing income documents, calculating maximum deductions, reviewing forms for inconsistencies, etc.) and tasks that are not automatable (meeting with new clients, explaining decisions to those clients, etc.). The OECD team then ran a probability model to find what percentage of jobs were at "high risk" (i.e., at least 70 percent of the tasks associated with the job could be automated). As noted, they found that in the United States only 9 percent of workers fell in the high-risk category. Applying that same model on twenty other OECD countries, the authors found that the percentage of high-risk jobs ranged from just 6 percent in Korea to 12 percent in Austria. Don't worry, the study seemed to say, reports of the death of work have been greatly exaggerated.

Unsurprisingly, that didn't settle the debate. The OECD's task-based approach came to hold sway among researchers, but not all of them agreed with the report's sanguine conclusions. In early 2017, researchers at PwC used the task-based approach to produce their own estimate, finding instead that 38 percent of jobs in the United States were at high risk of automation by the early 2030s. It was a striking divergence from the OECD's 9 percent, one that stemmed simply from using a slightly different algorithm in the calculations. Like the previous studies, the PwC authors are quick to note that this is merely an estimate of what jobs *could* be done by machines, and that actual job losses will be mitigated by regulatory, legal, and social dynamics.

After these wildly diverging estimates, researchers at the McKinsey Global Institute landed somewhere in the middle. I assisted the institute in its research related to China and coauthored a report with it on the Chinese digital landscape. Using the popular task-based approach, the McKinsey team estimated that around 50 percent of work tasks around the world are *already automatable.* For China, that number was pegged at 51.2 percent, with the United States coming in slightly lower, at 45.8 percent. But when it came to actual *job displacement,* the McKinsey researchers were less pessimistic. If there is rapid adoption of automation techniques (a scenario most comparable to the above estimates), 30 percent of work activities around the world could be automated by 2030, but only 14 percent of workers would need to change occupations.

So where does this survey of the literature leave us? Experts continue to be all over the map, with estimates of automation potential in the United States ranging from just 9 percent to 47 percent. Even if we stick to only the task-based approach, we still have a spread of 9 to 38 percent, a divide that could mean the difference between broad-based prosperity and an outright jobs crisis. That spread of estimates shouldn't cause us to throw up our hands in confusion. Instead, it should spur us to think critically about what these studies can teach us — and what they may have missed.

WHAT THE STUDIES MISSED

While I respect the expertise of the economists who pieced together the above estimates, I also respectfully disagree with the low-end estimates of the OECD. That difference is rooted in two disagreements: one in terms of the inputs of their equations, and one major difference in the way I envision AI disrupting labor markets. The quibble causes me to go with the higher-end estimates of PwC, and the difference in vision leads me to raise that number higher still.

My disagreement on inputs stems from the way the studies estimated the technical capabilities of machines in the years ahead. The 2013 Oxford study asked a group of machine-learning experts to predict whether seventy occupations would likely be automated in the coming two decades, using those assessments to project automatability more broadly. And though the OECD and PwC studies differed in how they divided up occupations and tasks, they basically stuck with the 2013 estimates of future capabilities.

Those estimates probably constituted the best guess of experts at the time, but significant advances in the accuracy and power of machine learning over the past five years have already moved the goal-posts. Experts back then may have been able to project some of the improvements that were on the horizon. But few, if any, experts predicted that deep learning was going to get *this good, this fast.* Those unexpected improvements are expanding the realm of the possible when it comes to real-world uses and thus job disruptions.

One of the clearest examples of these accelerating improvements is the ImageNet competition. In the competition, algorithms submitted by different teams are tasked with identifying thousands of different objects within millions of different images, such as birds, baseballs, screwdrivers, and mosques. It has quickly emerged as one of the most respected image-recognition contests and a clear benchmark for AI's progress in computer vision.

When the Oxford machine-learning experts made their estimates of technical capabilities in early 2013, the most recent Image-Net competition of 2012 had been the coming-out party for deep learning. Geoffrey Hinton's team used those techniques to achieve a record-setting error rate of around 16 percent, a large leap forward in a competition where no team had ever gotten below 25 percent.

That was enough to wake up much of the AI community to this thing called deep learning, but it was just a taste of what was to come. By 2017, almost every team had driven error rates below 5 percent — approximately the accuracy of humans performing the same task — with the average algorithm of that year making only one-third of the mistakes of the top algorithm of 2012. In the years since the Oxford experts made their predictions, computer vision has now surpassed human capabilities and dramatically expanded real-world use-cases for the technology.

Those amped-up capabilities extend far beyond computer vision. New algorithms constantly set and surpass records in fields like speech recognition, machine reading, and machine translation. While these strengthened capabilities don't constitute fundamental breakthroughs in AI, they do open the eyes and spark the imaginations of entrepreneurs. Taken together, these technical advances and emerging uses cause me to land on the higher end of task-based estimates, namely, PwC's prediction that 38 percent of U.S. jobs will be at high risk of automatability by the early 2030s.

TWO KINDS OF JOB LOSS: ONE-TO-ONE
REPLACEMENTS AND GROUND-UP DISRUPTIONS

But beyond that disagreement over methodology, I believe using only the task-based approach misses an entirely separate category of potential job losses: industry-wide disruptions due to new AI-empowered business models. Separate from the occupation- or task-based approach, I'll call this the *industry-based* approach.

Part of this difference in vision can be attributed to professional background. Many of the preceding studies were done by economists, whereas I am a technologist and early-stage investor. In predicting what jobs were at risk of automation, economists looked at what tasks a person completed while going about their job and asked whether a machine would be able to complete those same tasks. In other words, the task-based approach asked how possible it was to do a one-to-one replacement of a machine for a human worker.

My background trains me to approach the problem differently. Early in my career, I worked on turning cutting-edge AI technologies into useful products, and as a venture capitalist I fund and help build new startups. That work helps me see AI as forming two distinct threats to jobs: one-to-one replacements and ground-up disruptions.

Many of the AI companies I've invested in are looking to build a single AI-driven product that can replace a specific kind of worker — for instance, a robot that can do the lifting and carrying of a warehouse employee or an autonomous-vehicle algorithm that can complete the core tasks of a taxi driver. If successful, these companies will end up selling their products to companies, many of whom may lay off redundant workers as a result. These types of one-to-one replacements are exactly the job losses captured by economists using the task-based approach, and I take PwC's 38 percent estimate as a reasonable guess for this category.

But then there exists a completely different breed of AI startups: those that reimagine an industry from the ground up. These companies don't look to replace one human worker with one tailor-made robot that can handle the same tasks; rather, they look for new ways to satisfy the fundamental human need driving the industry.

Startups like Smart Finance (the AI-driven lender that employs no human loan officers), the employee-free F5 Future Store (a Chinese startup that creates a shopping experience comparable to the Amazon Go supermarket), or Toutiao (the algorithmic news app that employs no editors) are prime examples of these types of companies. Algorithms aren't displacing human workers at these companies, simply because the humans were never there to begin with. But as the lower costs and superior services of these companies drive gains to market share, they will apply pressure to their employee-heavy rivals. Those companies will be forced to adapt from the ground up — restructuring their workflows to leverage AI and reduce employees — or risk going out of business. Either way, the end result is the same: there will be fewer workers.

This type of AI-induced job loss is largely missing from the task-based estimates of the economists. If one applied the task-based approach to measuring the automatability of an editor at a news app, you would find dozens of tasks that can't be performed by machines. They can't read and understand news and feature articles, subjectively assess appropriateness for a particular app's audience, or communicate with reporters and other editors. But when Toutiao's founders built the app, they didn't look for an algorithm that could perform all of the above tasks. Instead, they reimagined how a news app could perform its core function — curate a feed of news stories that users want to read — and then did that by employing an AI algorithm.

I estimate this kind of from-the-ground-up disruption will affect about 10 percent of the workforce in the United States. The hardest hit industries will be those that involve high volumes of routine optimization work paired with external marketing or customer service: fast food, financial services, security, even radiology. These changes will eat away at employment in the "Human Veneer" quadrant of the earlier chart, with companies consolidating customer interaction tasks into a handful of employees, while algorithms do most of the grunt work behind the scenes. The result will be steep — though not total — reductions in jobs in these fields.

Putting together percentages for the two types of automatability — 38 percent from one-to-one replacements and about 10 percent from ground-up disruption — we are faced with a monumental challenge. Within ten to twenty years, I estimate we will be technically capable of automating 40 to 50 percent of jobs in the United States. For employees who are not outright replaced, increasing automation of their workload will continue to cut into their value-add for the company, reducing their bargaining power on wages and potentially leading to layoffs in the long term. We'll see a larger pool of unemployed workers competing for an even smaller pool of jobs, driving down wages and forcing many into part-time or "gig economy" work that lacks benefits.

This — and I cannot stress this enough — does *not* mean the country will be facing a 40 to 50 percent unemployment rate. Social frictions, regulatory restrictions, and plain old inertia will greatly slow down the actual rate of job losses. Plus, there will also be new jobs created along the way, positions that can offset a portion of these AI-induced losses, something that I explore in coming chapters. These could cut actual AI-induced net unemployment in half, to between 20 and 25 percent, or drive it even lower, down to just 10 to 20 percent.

These estimates are in line with those from the most recent research (as of this writing) that attempted to put a number on *actual job losses,* a February 2018 study by the consulting firm Bain and Company. Instead of wading into the minutiae of tasks and occupations, the Bain study took a macro-level approach, seeking to understand the interplay of three major forces acting on the global economy: demographics, automation, and inequality. Bain's analysis produced a startling bottom-line conclusion: by 2030, employers will need 20 to 25 percent fewer employees, a percentage that would equal 30 to 40 million displaced workers in the United States.

Bain acknowledged that some of these workers will be reabsorbed into new professions that barely exist today (such as robot repair technician), but predicted that this reabsorption would fail to make a meaningful dent in the massive and growing trend of displacement. And automation's impact will be felt far wider than even this 20 to 25 percent of displaced workers. The study calculated that if we include both displacement and wage suppression, a full *80 percent* of all workers will be affected.

This would constitute a devastating blow to working families. Worse still, this would not be a temporary shock, like the fleeting brush with 10 percent unemployment that the United States experienced following the 2008 financial crisis. Instead, if left unchecked, it could constitute the new normal: an age of full employment for intelligent machines and enduring stagnation for the average worker.

U.S.-CHINA COMPARISON: MORAVEC'S REVENGE

But what about China? How will its workers fare in this brave new economy? Few good studies have been conducted on the impacts of automation here, but the conventional wisdom holds that Chinese people will be hit much harder, with intelligent robots spelling the end of a golden era for workers in the "factory of the world." This prediction is based on the makeup of China's workforce, as well as a gut-level intuition about what kinds of jobs become automated.

Over one-quarter of Chinese workers are still on farms, with another quarter involved in industrial production. That compares with less than 2 percent of Americans in agriculture and around 18 percent in industrial jobs. Pundits such as *Rise of the Robots* author Martin Ford have argued that this large base of routine manual labor could make China "ground zero for the economic and social disruption brought on by the rise of the robots." Influential technology commentator Vivek Wadhwa has similarly predicted that intelligent robotics will erode China's labor advantage and bring manufacturing back to the United States en masse, albeit without the accompanying jobs for humans. "American robots work as hard as Chinese robots," he wrote, "and they also don't complain or join labor unions."

These predictions are understandable given the recent history of automation. Looking back at the last hundred years of economic evolution, blue-collar workers and farmhands have faced the steepest job losses from physical automation. Industrial and agricultural tools (think forklifts and tractors) greatly increased the productivity of each manual laborer, reducing demand for workers in these sectors. Projecting this same transition out into the age of AI, the conventional wisdom views China's farm and factory laborers as caught squarely in the crosshairs of intelligent automation. In contrast, America's heavily service-oriented and white-collar economy has a greater buffer against potential job losses, protected by college degrees and six-figure incomes.

In my opinion, the conventional wisdom on this is backward. While China will face a wrenching labor-market transition due to automation, large segments of that transition may arrive later or move slower than the job losses wracking the American economy. While the simplest and most routine factory jobs — quality control and simple assembly-line tasks — will likely be automated in the coming years, the remainder of these manual labor tasks will be tougher for robots to take over. This is because the intelligent automation of the twenty-first century operates differently than the physical automation of the twentieth century. Put simply, it's far easier to build AI algorithms than to build intelligent robots.

Core to this logic is a tenet of artificial intelligence known as Moravec's Paradox. Hans Moravec was a professor of mine at Carnegie Mellon University, and his work on artificial intelligence and robotics led him to a fundamental truth about combining the two: contrary to popular assumptions, it is relatively easy for AI to mimic the high-level intellectual or computational abilities of an adult, but it's far harder to give a robot the perception and sensorimotor skills of a toddler. Algorithms can blow humans out of the water when it comes to making predictions based on data, but robots still can't perform the cleaning duties of a hotel maid. In essence, AI is great at thinking, but robots are bad at moving their fingers.

Moravec's Paradox was articulated in the 1980s, and some things have changed since then. The arrival of deep learning has provided machines with superhuman perceptual abilities when it comes to voice or visual recognition. Those same machine-learning breakthroughs have also turbocharged the intellectual abilities of machines, namely, the power of spotting patterns in data and making decisions. But the fine motor skills of robots — the ability to grasp and manipulate objects — still lag far behind humans. While AI can beat the best humans at Go and diagnose cancer with extreme accuracy, it cannot yet appreciate a good joke.

THE ASCENT OF THE ALGORITHMS
AND RISE OF THE ROBOTS

This hard reality about algorithms and robots will have profound effects on the *sequence* of AI-induced job losses. The physical automation of the past century largely hurt blue-collar workers, but the coming decades of intelligent automation will hit white-collar workers first. The truth is that these workers have far more to fear from the algorithms that exist today than from the robots that still need to be invented.

In short, AI algorithms will be to many white-collar workers what tractors were to farmhands: a tool that dramatically increases the productivity of each worker and thus shrinks the total number of employees required. And unlike tractors, algorithms can be shipped instantly around the world at no additional cost to their creator. Once that software has been sent out to its millions of users — tax-preparation companies, climate-change labs, law firms — it can be constantly updated and improved with no need to create a new physical product.

Robotics, however, is much more difficult. It requires a delicate interplay of mechanical engineering, perception AI, and fine-motor manipulation. These are all solvable problems, but not at nearly the speed at which pure software is being built to handle white-collar cognitive tasks. Once that robot is built, it must also be tested, sold, shipped, installed, and maintained on-site. Adjustments to the robot's underlying algorithms can sometimes be made remotely, but any mechanical hiccups require hands-on work with the machine. All these frictions will slow down the pace of robotic automation.

This is not to say that China's manual laborers are safe. Drones for deploying pesticides on farms, warehouse robots for unpacking trucks, and vision-enabled robots for factory quality control will all dramatically reduce the jobs in these sectors. And Chinese companies are indeed investing heavily in all of the above. The country is already the world's top market for robots, buying nearly as many as Europe and the Americas combined. Chinese CEOs and political leaders are united in pushing for the steady automation of many Chinese factories and farms.

But the resulting blue-collar job losses in China will be more gradual and piecemeal than the sweeping impact of algorithms on white-collar workers. While the right digital algorithm can hit like a missile strike on cognitive labor, robotics' assault on manual labor is closer to trench warfare. Over the long term, I believe the number of jobs at risk of automation will be similar for China and the United States. American education's greater emphasis on creativity and interpersonal skills may give it an employment edge on a long enough time scale. However, when it comes to adapting to these changes, speed matters, and China's particular economic structure will buy it some time.

THE AI SUPERPOWERS VERSUS ALL THE REST

Whatever gaps exist between China and the United States, those differences will pale in comparison between these two AI superpowers and the rest of the world. Silicon Valley entrepreneurs love to describe their products as "democratizing access," "connecting people," and, of course, "making the world a better place." That vision of technology as a cure-all for global inequality has always been something of a wistful mirage, but in the age of AI it could turn into something far more dangerous. If left unchecked, AI will dramatically exacerbate inequality on both international and domestic levels. It will drive a wedge between the AI superpowers and the rest of the world, and may divide society along class lines that mimic the dystopian science fiction of Hao Jingfang.

As a technology and an industry, AI naturally gravitates toward monopolies. Its reliance on data for improvement creates a self-perpetuating cycle: better products lead to more users, those users lead to more data, and that data leads to even better products, and thus more users and data. Once a company has jumped out to an early lead, this kind of ongoing repeating cycle can turn that lead into an insurmountable barrier to entry for other firms.

Chinese and American companies have already kick-started this process, leaping out to massive leads over the rest of the world. Canada, the United Kingdom, France, and a few other countries play host to top-notch talent and research labs, but they often lack the other ingredients needed to become true AI superpowers: a large base of users and a vibrant entrepreneurial and venture-capital ecosystem. Other than London's DeepMind, we have yet to see groundbreaking AI companies emerge from these countries. All of the seven AI giants and an overwhelming portion of the best AI engineers are already concentrated in the United States and China. They are building huge stores of data that are feeding into a variety of different product verticals, such as self-driving cars, language translation, autonomous drones, facial recognition, natural-language processing, and much more. The more data these companies accumulate, the harder it will be for companies in any other countries to ever compete.

As AI spreads its tentacles into every aspect of economic life, the benefits will flow to these bastions of data and AI talent. PwC estimates that the United States and China are set to capture a full 70 percent of the $15.7 trillion that AI will add to the global economy by 2030, with China alone taking home $7 trillion. Other countries will be left to pick up the scraps, while these AI superpowers will boost productivity at home and harvest profits from markets around the globe. American companies will likely lay claim to many developed markets, and China's AI juggernauts will have a better shot at winning over Southeast Asia, Africa, and the Middle East.

I fear this process will exacerbate and significantly grow the divide between the AI haves and have-nots. While AI-rich countries rake in astounding profits, countries that haven't crossed a certain technological and economic threshold will find themselves slipping backward and falling farther behind. With manufacturing and services increasingly done by intelligent machines located in the AI superpowers, developing countries will lose the one competitive edge that their predecessors used to kick-start development: low-wage factory labor.

Large populations of young people used to be these countries' greatest strengths. But in the age of AI, that group will be made up of displaced workers unable to find economically productive work. This sea change will transform them from an engine of growth to a liability on the public ledger — and a potentially explosive one if their governments prove unable to meet their demands for a better life.

Deprived of the chance to claw their way out of poverty, poor countries will stagnate while the AI superpowers take off. I fear this ever-growing economic divide will force poor countries into a state of near-total dependence and subservience. Their governments may try to negotiate with the superpower that supplies their AI technology, trading market and data access for guarantees of economic aid for their population. Whatever bargain is struck, it will not be one based on agency or equality between nations.

THE AI INEQUALITY MACHINE

The same push toward polarization playing out across the global economy will also exacerbate inequality within the AI superpowers. AI's natural affinity for monopolies will bring winner-take-all economics to dozens more industries, and the technology's skill biases will generate a bifurcated job market that squeezes out the middle class. The "great decoupling" of productivity and wages has already created a tear between the 1 percent and the 99 percent. Left to its own devices, artificial intelligence, I worry, will take this tear and rip it wide open.

We already see this trend toward monopolization in the online world. The internet was supposed to be a place of freewheeling competition and a level playing field, but in a few short years many core online functions have turned into monopolistic empires. For much of the developed world, Google rules search engines, Facebook dominates social networks, and Amazon owns e-commerce. Chinese internet companies tend to worry less about "staying in their lane," so there are more skirmishes between these giants, but the vast majority of China's online activity is still funneled through just a handful of companies.

AI will bring that same monopolistic tendency to dozens of industries, eroding the competitive mechanisms of markets in the process. We could see the rapid emergence of a new corporate oligarchy, a class of AI-powered industry champions whose data edge over the competition feeds on itself until they are entirely untouchable. American antitrust laws are often difficult to enforce in this situation, because of the requirement in U.S. law that plaintiffs prove the monopoly is actually harming consumers. AI monopolists, by contrast, would likely be delivering better and better services at cheaper prices to consumers, a move made possible by the incredible productivity and efficiency gains of the technology.

But while these AI monopolies drive down prices, they will also drive up inequality. Corporate profits will explode, showering wealth on the elite executives and engineers lucky enough to get in on the action. Just imagine: How profitable would Uber be if it had no drivers? Or Apple if it didn't need factory workers to make iPhones? Or Walmart if it paid no cashiers, warehouse employees, and truck drivers?

Driving income inequality will be the emergence of an increasingly bifurcated labor market. The jobs that do remain will tend to be either lucrative work for top performers or low-paying jobs in tough industries. The risk of replacement cited in the earlier figures reflects this. The most difficult jobs to automate — those in the top-right corner of the "Safe Zone" — include both ends of the income spectrum: CEOs and healthcare aides, venture capitalists and masseuses.

Meanwhile, many of the professions that form the bedrock of the middle class — truck drivers, accountants, office managers — will be hollowed out. Sure, we could try to transition these workers into some of the highly social, highly dexterous occupations that will remain safe. Home healthcare aide, techno-optimists point out, is the fastest-growing profession in America. But it's also one of the lowest paid, with an annual salary of around $22,000. A rush of newly displaced workers trying to enter the industry will only exert more downward pressure on that number.

Pushing more people into these jobs while the rich leverage AI for huge gains doesn't just create a society that is dramatically unequal. I fear it will also prove unsustainable and frighteningly unstable.

A GRIM PICTURE

When we scan the economic horizon, we see that artificial intelligence promises to produce wealth on a scale never before seen in human history—something that should be a cause for celebration. But if left to its own devices, AI will also produce a global distribution of wealth that is not just more unequal but hopelessly so. AI-poor countries will find themselves unable to get a grip on the ladder of economic development, relegated to permanent subservient status. AI-rich countries will amass great wealth but also witness the widespread monopolization of the economy and a labor market divided into economic castes.

Make no mistake: this is not just the normal churn of capitalism's creative destruction, a process that has previously helped lead to a new equilibrium of more jobs, higher wages, and a better quality of life for all. The free market is supposed to be self-correcting, but these self-correcting mechanisms break down in an economy driven by artificial intelligence. Low-cost labor provides no edge over machines, and data-driven monopolies are forever self-reinforcing.

These forces are combining to create a unique historical phenomenon, one that will shake the foundations of our labor markets, economies, and societies. Even if the most dire predictions of job losses don't fully materialize, the social impact of wrenching inequality could be just as traumatic. We may never build the folding cities of Hao Jingfang's science fiction, but AI risks creating a twenty-first-century caste system, one that divides the population into the AI elite and what historian Yuval N. Harari has crudely called the "useless class," people who can never generate enough economic value to support themselves. Even worse, recent history has shown us just how fragile our political institutions and social fabric can be in the face of intractable inequality. I fear that recent upheavals are only a dry run for the disruptions to come in the age of AI.

The resulting turmoil will take on political, economic, and social dimensions, but it will also be intensely personal. In the centuries since the Industrial Revolution, we have increasingly come to see our work not just as a means of survival but as a source of personal pride, identity, and real-life meaning. Asked to introduce ourselves or others in a social setting, a job is often the first thing we mention. It fills our days and provides a sense of routine and a source of human connections. A regular paycheck has become a way not just of rewarding labor but also of signaling to people that one is a valued member of society, a contributor to a common project.

Severing these ties — or forcing people into downwardly mobile careers — will damage so much more than our financial lives. It will constitute a direct assault on our sense of identity and purpose. Speaking to the *New York Times* in 2014, a laid-off electrician named Frank Walsh described the psychological toll of intractable unemployment.

"I lost my sense of worth, you know what I mean?" Walsh observed. "Somebody asks you 'What do you do?' and I would say, 'I'm an electrician.' But now I say nothing. I'm not an electrician anymore."

That loss of meaning and purpose has very real and serious consequences. Rates of depression triple among those unemployed for six months, and people looking for work are twice as likely to commit suicide as the gainfully employed. Alcohol abuse and opioid overdoses both rise alongside unemployment rates, with some scholars attributing rising mortality rates among uneducated white Americans to declining economic outcomes, a phenomenon they call "deaths of despair."

The psychological damage of AI-induced unemployment will cut even deeper. People will face the prospect of not just being temporarily out of work but of being permanently excluded from the functioning of the economy. They will watch as algorithms and robots easily outperform them at tasks and skills they spent their whole lives mastering. It will lead to a crushing feeling of futility, a sense of having become obsolete in one's own skin.

The winners of this AI economy will marvel at the awesome power of these machines. But the rest of humankind will be left to grapple with a far deeper question: when machines can do everything that we can, what does it mean to be human?

That's a question that I found myself grappling with in the depths of my own personal crisis of mortality and meaning. That crisis brought me to a very dark place, one that pushed my body to the limit and challenged my deepest-held assumptions about what matters in life. But it was that process — and that pain — that opened my eyes to an alternate ending to the story of human beings and artificial intelligence.

7

★

THE WISDOM OF CANCER

The profound questions raised by our AI future — questions about the relationship among work, value, and what it means to be human — hit close to home for me.

For most of my adult life, I have been driven by an almost fanatical work ethic. I gave nearly all my time and energy to my job, leaving very little for family or friends. My sense of self-worth was derived from my achievements at work, from my ability to create economic value and to expand my own influence in the world.

I had spent my research career working to build ever more powerful artificial intelligence algorithms. In doing this, I came to view my own life as a kind of optimization algorithm with a clear goals: maximize personal influence and minimize anything that doesn't contribute to that goal. I sought to quantify everything in my life, balancing these "inputs" and fine-tuning the algorithm.

I didn't entirely neglect my wife or daughters, but I always sought to spend *just enough* time with them so they didn't complain. As soon as I felt I had met that bar, I would race back to work, answering emails, launching products, funding companies, and making speeches. Even in the depths of sleep, my body would naturally wake itself up twice each night — at 2 a.m. and 5 a.m. — to reply to emails from the United States.

That obsessive dedication to work did not go unrewarded. I became one of the top AI researchers in the world, founded the best computer science research institute in Asia, started Google China, created my own successful venture-capital fund, wrote multiple best-selling books in Chinese, and amassed one of the largest so-cial media followings in China. By any objective metric, my so-called personal algorithm was a smashing success.

And then things came to a grinding halt.

In September 2013, I was diagnosed with stage IV lymphoma. In an instant, my world of mental algorithms and personal achieve-ments came crashing down. None of those things could save me now, or give me comfort and a sense of meaning. Like so many peo-ple forced to suddenly face their own mortality, I was filled with fear for my future and with a deep, soul-aching regret over the way I had lived my life.

Year after year, I had ignored the opportunity to spend time and share love with the people closest to me. My family had given me nothing but warmth and love, and I had responded to that on the ba-sis of cold calculations. In effect, mesmerized by my quest to create machines that thought like people, I had turned into a person that thought like a machine.

My cancer would go into remission, sparing my life, but the epiphanies sparked by this personal confrontation with death have stuck with me. They've led me to reshuffle my priorities and to to-tally change my life. I spend far more time with my wife and daugh-ters, and moved to be closer to my aging mother. I have dramati-cally cut down my presence on social media, pouring that time into meeting with and trying to help young people who reach out to me. I've asked for forgiveness from those I have wronged and sought to be a kinder and more empathetic coworker. Most of all, I've stopped viewing my life as an algorithm that optimizes for influence. Instead, I try to spend my energy doing the one thing I've found that truly brings meaning to a person's life: sharing love with those around us.

This near-death experience also gave me a new vision for how humans can coexist with artificial intelligence. Yes, this technology will both create enormous economic value and destroy an astound-ing number of jobs. If we remain trapped in a mindset that equates our economic value with our worth as human beings, this transition to the age of AI will devastate our societies and wreak havoc on our individual psychologies.

But there is another path, an opportunity to use artificial intelligence to double down on what makes us truly human. This path won't be easy, but I believe it represents our best hope of not just surviving in the age of AI but actually thriving. It's a journey that I've taken in my own life, one that turned my focus from machines back to people, and from intelligence back to love.

DECEMBER 16, 1991

The routinized chaos of childbirth swirled all around me. Nurses and doctors in sanitary scrubs streamed in and out of the room, checking measurements and swapping out IV drips. My wife, Shen-Ling, lay on the hospital bed, fighting through the most physically and mentally draining act that a human being can perform: bringing another human into the world. It was December 16, 1991, and I was about to become a father for the first time.

Our attending doctor told me it was going to be a complex labor because the baby was in the sunny-side up position, with her head facing toward the belly instead of toward the back. That meant Shen-Ling might require a cesarean section. I paced the room anxiously, even more on edge than most expectant fathers on the big day. I was worried about Shen-Ling and the baby's health, but my mind wasn't entirely in that delivery room.

That's because this was the day I was scheduled to deliver a presentation to John Sculley, my CEO at Apple and one of the most powerful men in the technology world. A year earlier, I had joined Apple as the chief scientist for speech recognition, and this presentation was my chance to win Sculley's endorsement for our proposal to include speech synthesis in every Macintosh computer and speech recognition in all new types of Macs.

My wife's labor continued, and I kept checking the clock. I desperately hoped that she would have the baby in time for me to be there for the birth and also make it back to headquarters in time for the meeting. As I paced the room, my coworkers called and asked if we should cancel the meeting or perhaps have my lieutenant give the presentation to Sculley.

"No," I told them. "I think I can make it."

But as the labor dragged on, it was looking increasingly unlikely that this would happen, and I was genuinely torn about what I should do: stay by my wife's side or rush off to an important meeting. Presented with a "problem" like this, my well-trained engineering mind kicked into high gear. I weighed all options in terms of inputs and outputs, maximizing my impact on measurable results.

Witnessing the birth of my first child would be great, but my daughter would be born whether I was there or not. On the other hand, if I missed this presentation to Sculley, it could have a very substantial and quantifiable impact. Maybe the software wouldn't respond well to my replacement's voice — I had a knack for coaxing the best performance out of it — and Sculley might shelve speech-recognition research indefinitely. Or maybe he would greenlight the project but then place someone else in charge of it. I imagined that the fate of artificial intelligence research hung in the balance, and maximizing the chances of success simply meant I had to be in that room for the presentation.

I was in the midst of these mental calculations when the doctor informed me that they would be performing an immediate cesarean section. My wife was rushed off to an operating room with me in tow, and within an hour Shen-Ling and I were holding our baby daughter. We all had some time together, and with little time left to spare, I took off for the presentation.

It went extremely well. Sculley both greenlighted the project and demanded a full-on publicity campaign around what I had created. That campaign led to a high-profile TED talk, write-ups in the *Wall Street Journal,* and an appearance on *Good Morning America* in 1992, with John Sculley and I demonstrating the technology for millions of viewers. On the program, we used voice commands to schedule an appointment, write a check, and program a VCR, showcasing the earliest examples of futuristic functions that wouldn't go mainstream for another twenty years, with Apple's Siri and Amazon's Alexa. These triumphs filled me with great personal pride and also turbocharged my career.

But looking back, it's not those career successes that stick in my mind. It's the scene in that hospital room. If I had been forced to choose between the birth of my first child and that Apple meeting, I likely would have chosen the meeting.

Today, I must confess that I find this deeply embarrassing but not entirely baffling. That's because this wasn't just about one meeting. It was a manifestation of the machine-like mentality that had dominated my life for decades.

THE IRONMAN

As a young man, computer science and artificial intelligence resonated with me because the crystal logic of the algorithms mirrored my own way of thinking. At the time, I processed everything in my life — friendships, work, and family time — as variables or inputs in my own mental algorithm. They were things to be quantified and metered out in the precise amounts required to achieve a specific outcome.

Like any good algorithm, I of course had to balance multiple goals. Self-driving cars don't just optimize for getting you home as fast as possible; they must do so without breaking any laws and while minimizing the risk of accidents. Likewise, I had to make certain tradeoffs between my personal and professional lives. I hadn't been a completely absent father, neglectful husband (the episode of my daughter's birth notwithstanding), or ungrateful son. My social algorithms were good enough that I made a point of remembering anniversaries, giving thoughtful gifts, and spending some time with the people in my family.

But I approached these as minimization functions, looking for ways to achieve the desired result while putting in the least amount of time possible. I always weighted the master algorithm heavily in favor of my own career goals to maximize time at work, personal influence, and status within my profession.

When I was given vacations of four weeks, I would spend one or two weeks with my mother in Taiwan or with my family in Beijing and then head right back to work. Even when a surgical procedure forced me to remain lying flat in bed for two weeks, I couldn't let my work go. I had a metal crane built that suspended a computer monitor above my pillow and connected it with a keyboard and mouse that I could lay across my lap. I was back to answering emails within hours of the surgery.

I wanted my employees, bosses, and fans to see me as a super-charged productivity machine, someone who did twice the work and needed half the rest of a normal human being. It also gave my team the not-so-subtle suggestion that I expected similar effort from them. My coworkers started calling me by the nickname "Ironman," and I loved it.

That work ethic powered an exhilarating lifestyle. I had a chance to stand at the frontier of science, the peak of global business, and in the limelight of national celebrity. In 2013, I was honored as one of the *Time* 100, the magazine's list of the most influential people in the world.

WHAT DO YOU WANT ON YOUR TOMBSTONE?

Each of those achievements just added more fuel to my internal fire. They pushed me to work harder and to preach this lifestyle to millions of young Chinese people. I wrote best-selling books with titles like *Be Your Personal Best* and *Making a World of Difference*. I traveled to college campuses around the country to deliver inspirational speeches. China was reemerging as a global power after centuries of poverty, and I exhorted Chinese students to seize the moment and make their own mark on history.

Ironically, I concluded these lectures with a striking image: a picture of my own tombstone. I told them that the best way to find one's calling was to picture your own grave and imagine what you want written on it. I said that my mission was clear, and my tombstone was ready:

> Here lies Kai-Fu Lee,
> scientist and business executive.
> Through his work at top technology companies
> he turned complex technical advances into products
> that everyone could use
> and everyone could benefit from.

It made for a fantastic conclusion to the speeches, a call to action that resonated with the ambition pulsing through the country at the time. China was evolving and growing as fast as any country in his-

tory, and the excitement was palpable. I felt perfectly in my element
and at the height of my powers.

After leaving Google and founding Sinovation Ventures, I began
to spend more time mentoring young people. I used my massive following on the Twitter-like platform Weibo to engage directly with
Chinese students, offering them guidance and writing open letters
that were collected into books. Although I remained the head of one
of the country's most prestigious venture-capital funds, students began referring to me as "Teacher Kai-Fu," an honorific that in China
combined great respect and also a certain closeness.

I basked in this role as a mentor to millions of students. I believed
that this turn toward "teaching" proved my own selflessness and
genuine desire to help others. In my speeches at Chinese universities, I kept the tombstone portion but changed the epitaph:

> Here lies Kai-Fu Lee,
> who had a love for education
> during the time of China's rise.
> Through writing, the internet, and lectures,
> he helped many young students,
> who lovingly called him "Teacher Kai-Fu."

Delivering that speech to enraptured audiences gave me a rush.
The new epitaph made for an even better ending, I thought, speaking to my substantial influence and also a certain wisdom that came
with age. I had gone from scientist to engineer, and from executive
to teacher. Along the way, I had managed to maximize my impact on
the world while giving my fans a sense of warmth and empathy. The
algorithm of my mind, I told myself, had been tuned to perfection.

It would take an encounter with the reality that lay behind that
tombstone — my own mortality — to understand just how foolish and
misguided my calculations had been.

DIAGNOSIS

The technician in charge of the PET scan was all business. After he
showed me into the room, he immediately set about inputting my
information and then programming the imaging device. Each year

my wife and I traveled back to Taiwan for our medical checkups. Earlier in 2013, one of our close relatives had been diagnosed with cancer, and so my wife decided that this year we would both get MRI and CT scans. After our checkup, my doctor said that he'd found something during the preliminary scans, and that I should come back in for a PET scan.

While MRI and CT scans require an expert eye to decipher, the results of a PET scan are relatively easy for anyone to understand. Patients are injected with a radioactive tracer, a dose of glucose that contains a tiny amount of a radioisotope. Cancerous cells tend to absorb sugar more intensely than other parts of the body, so these radioisotopes will tend to cluster around potentially cancerous growths. Computer images generated by the scans represent those clusters in bright red. Before we began, I asked the technician if I could see the scan once I was finished.

"I'm not a radiologist," he said. "But yes, I can show you the pictures."

With that, I lay down on the machine and disappeared into the circular tube within. When I emerged forty-five minutes later, the technician was still hunched over his computer, staring intently at the screen and clicking his mouse in rapid succession.

"Can I see the pictures now?" I asked.

"You really should go see your radiologist first," he replied without looking up.

"But you told me that I could see it," I protested. "It's right on the screen there, isn't it?"

Giving in to my insistence, he pivoted the computer monitor around to face me. A cold chill seized my chest, turning into an icy shiver as it spread across my skin. The black scan of my body was dotted with numerous red blotches across my stomach and abdomen.

"What are all these red things?" I said, my jaw beginning to quiver.

The technician wouldn't look me in the eye. I felt that initial chill turning into a hot panic.

"Are these tumors?" I demanded.

"There's a probability that these are tumors," he replied, still not making eye contact. "But you should really stay calm and go see your radiologist."

My mind was swimming, but my body continued on autopilot. I asked the technician to please print the scan for me, and I headed down the hall to the radiologist's office. I didn't have an appointment with the radiologist yet, and it was against the rules for them to examine my printouts casually, but I begged and pleaded until someone there agreed to make an exception. After looking over the scans, the radiologist told me that the pattern of these clusters meant that I had lymphoma. When I asked what stage it was in, he tried to deflect the question.

"Well, it's complex. We have to find out what kind —"

I cut him off: "But what stage is it?"

"Probably stage four."

I walked out of the room and then the hospital clutching the paper with both hands, holding it close to my chest so no one passing by could glimpse what was growing inside me. I decided I had to go home and write my will.

THE WILL

That teardrop on the page was going to cost me an hour of hard work. I had tried to dab it away with tissue as it grew heavy on my eyelash, but I was a second too late and it dropped to the paper below, landing squarely atop the Chinese character for "Lee." As the salty tear mixed with the ink on the page, it formed a tiny black puddle that slowly seeped into the paper. I had to start over.

For a will to be in effect immediately in Taiwan, it must be handwritten, with no blemishes or corrections. It's a straightforward requirement, if a bit dated. To accomplish this, I took out my best ink pen, the same one I'd used to sign hundreds of copies of the books I had written: a best-selling autobiography and several volumes encouraging young Chinese people to take control of their careers through hard work. That pen was failing me now. My hand quivered with anxiety, and my mind couldn't shake the image of that PET scan. I tried to remain focused on the lawyer's instructions for the will, but as my mind wandered, my pen would slip, marring one Chinese character and forcing me to start from scratch.

It wasn't just the memory of those fiery red blotches that made writing so difficult. My will had to be written in the traditional Chinese characters used in Taiwan — complex combinations of strokes, hooks, and flourishes far more intricate and elegant than the simplified characters used in mainland China. These characters constitute one of the oldest written languages still in use today, and I'd grown up immersed in it. I devoured epic kung-fu novels as a kid and even wrote one of my own when I was in elementary school.

At the age of eleven I moved from Taiwan to Tennessee, a move inspired by my older brother, who was working in the United States and told my mother that Taiwan's education system was too rigid and exam-oriented for a kid like me. It was tough for my mother to watch as her baby boy moved halfway around the world, and when we said goodbye, she made me promise one thing: that I would write her a letter in Chinese each week. In her letters back to me, she included a copy of the last letter I had sent to her, with corrections to those characters I had written wrong. That correspondence kept the written Chinese language alive for me as I went through high school, college, and graduate school in the United States.

As I threw myself into a prestigious job at Apple in the early 1990s, our handwritten correspondence grew less frequent. When I moved to Beijing and began work with Microsoft, computers ate away more and more of the time I'd spent crafting traditional characters by hand. Writing Chinese on a computer was easier; it required typing out the romanized spelling of a Chinese word (for example, *nihao*) and then selecting the corresponding characters from a list. Artificial intelligence has further streamlined the process by predicting and automatically selecting the characters based on context. That technology has made typing Chinese almost as efficient as hammering out alphabetic languages like English.

But gains in efficiency had turned into losses of memory. As I now sat hunched over the paper, I struggled to summon the shape of the characters after decades of neglect. I kept forgetting a dot or adding a horizontal stroke where it wasn't meant to be. Each time I fudged a character, I would crumple up the paper and begin again.

My will was just a page long, and in it I left everything to my wife, Shen-Ling. But my lawyer insisted that I write out four copies of that one page, each one to account for a different possible contingency. What if Shen-Ling died before me? Then I would give it all to my two daughters. What if one of them died? What if Shen-Ling and both of them died? It is an absurd set of hypotheticals to foist on someone grappling with his own mortality, but the law doesn't carve out exceptions for a person's internal distress.

Those hypotheticals did, however, refocus my mind on what mattered. Not the management of my financial assets but the people in my life. Ever since I saw that PET scan, the world had seemed to dissolve into a whirlpool of despair, one with me at the center. Why did this happen to me? I'd never intentionally hurt anyone. I had always tried to make the world a better place, to create technologies that made life easier for people. I had used my fame in China to educate and inspire young people. I had done nothing to deserve dying at the age of fifty-three.

Every one of those thoughts began with "I" and centered on self-righteous assertions of my own "objective" value. It wasn't until I wrote down the names of my wife and daughters, character by character in black ink, that I snapped out of this egocentric wallowing and self-pity. The real tragedy wasn't that I might not live much longer. It was that I had lived so long without generously sharing love with those so close to me.

Seeing my ultimate end point threw my life into sharp focus and turned my egocentric wallowing inside out. I stopped asking why the world had done this to me, or lamenting that all my achievements couldn't save me now. I began asking new questions: Why had I wanted so desperately to turn myself into a productivity machine? Why hadn't I taken the time to share love with others? Why did I ignore the very essence that made me human?

LIVING TOWARD DEATH

As the sun set on Taipei, I sat alone at the table, looking at the four copies of my will, which had taken me four hours to write. My wife was in Beijing with our younger daughter, and I sat alone in the living room of my mother's home. In the next room, my mother was lying down. She had for years suffered from dementia, and while she could still recognize her son, she had little ability to understand the world around her.

For a moment, I felt grateful for the illness that clouded her mind — if she could understand the diagnosis that had just been delivered, I feared it would have broken her. She had given birth to me when she was forty-four, an age at which doctors urged her not to go through with the pregnancy. She refused to entertain that idea, seeing the pregnancy through and then showering me with endless affection. I was her baby, and she loved nothing more than feeding me her handmade spicy Sichuan dumplings, delicately wrapped bundles of pork that practically melted on your tongue.

When I made the move to Tennessee, despite not speaking a word of English, my mother came and stayed with me for my first six months in America, just to make sure I was all right. Preparing to return home to Taiwan, she asked only that I continue to write her those letters in Chinese each week, a way to keep me close to her heart and rooted in the culture of my ancestors.

She was someone who had spent her whole life sharing love with her children. Sitting at her dining table while she lay in the next room, I was racked by wave after wave of remorse. How had I been raised by such an emotionally generous woman and yet lived my life so focused on myself? Why had I never told my father that I loved him? Or truly shown the depth of caring for my mother before the dementia took hold?

The hardest thing about facing death isn't the experiences you won't get to have. It's the ones you can't have back. Palliative care nurse and author Bronnie Ware has written extensively on the most common regrets that her terminally ill patients expressed in their final weeks of life. Facing the ultimate, these patients were able to look back on their lives with a clarity that escapes those of us absorbed in our daily grind. They spoke of the pain of not having lived a life true to themselves, the regret at having focused so obsessively on their work, and the realization that it's the people in your life who give it true meaning. None of these people looked back on their lives wishing they had worked harder, but many of them found themselves wishing they had spent more time with the ones they loved.

"It all comes down to love and relationships in the end," Ware wrote in the blog post that launched her book. "That is all that remains in the final weeks: love and relationships."

Sitting at my mother's table, this simple truth now burned within me. My mind swam backward through time, dipping in and out of memories of my daughters, my wife, and my parents. I hadn't ignored the relationships in my life; on the contrary, I had very precisely accounted for each one. I had quantified them all and calculated the optimal allocation of time needed to achieve my objectives. Now I felt a gaping sense of emptiness, of irretrievable loss, about how little time for loved ones my mental algorithm had deemed "optimal." This algorithmic way of thinking wasn't just "suboptimal" at allocating time. It was robbing me of my own humanity.

THE MASTER ON THE MOUNTAIN

Like any epiphany worth having, these thoughts took time to truly sink in. I had felt something shift within me, but it would require patience and brutally honest self-examination to turn these pangs of regret into a new way of engaging with the world around me.

Soon after my diagnosis, a friend recommended I visit the Fo Guang Shan Buddhist monastery in the south of Taiwan. Venerable Master Hsing Yun, a rotund monk with a soft smile, founded Fo Guang Shan in 1967 and remains at the monastery today. His monastic order practices what is called "humanistic Buddhism," a modern approach to the faith that seeks to integrate core practices and precepts into our daily lives. Its monks eschew the stern mysteriousness of traditional Buddhism, instead embracing life with unconcealed joy. The monastery welcomes visitors from all backgrounds, sharing with them simple practices and gentle wisdom. Around the monastery, you see couples getting married, monks enjoying a good laugh, and tourists taking a moment out of their busy lives to bask in the calm exuding from the people there.

I had practiced Christianity while growing up in the United

States, and although I no longer ascribe to a religious faith, I maintain a belief in a creator of this world and a power greater than our own. In visiting the monastery, I didn't have any particular ambition —just a desire to spend a few days meditating on what I was experiencing, and reflecting on the life I had lived.

One day after early morning classes, I was asked to join Master Hsing Yun for a vegetarian breakfast. The sun had not yet risen as we ate multigrain bread, tofu, and porridge. Master Hsing Yun now uses a wheelchair to get around, but his mind remains clear and sharp. Partway through our meal, he turned to me with a blunt question.

"Kai-Fu, have you ever thought about what your goal is in life?"

Without thinking, I reflexively gave him the answer I had given to myself and others for decades: "To maximize my impact and change the world."

Speaking those words, I felt the burning embarrassment that comes when we expose our naked ambitions to others. The feeling was magnified by the silence emanating from the monk across the table. But my answer was an honest one. This quest to maximize my impact was like a tumor that had always lived inside of me, ever tenacious and always growing. I had read widely in philosophy and religious texts, but for decades had never critically examined or doubted this core motivating belief within me.

For a moment, Master Hsing Yun said nothing, using a piece of bread to wipe the last scraps of breakfast from his wooden bowl. I shifted uncomfortably in my seat.

"What does it really mean to 'maximize impact'?" he began. "When people speak in this way, it's often nothing but a thin disguise for ego, for vanity. If you truly look within yourself, can you say for sure that what motivates you is not ego? It's a question you must ask your own heart, and whatever you do, don't try to lie to yourself."

My mind raced with rebuttals. I searched for the airtight logic that would redeem my actions. The days since my diagnosis had been an agonizing exercise in regret about the way I had engaged with my family and friends. I was slowly coming to terms with the emptiness of my emotional life. But as described in Elisabeth Kübler-Ross's theory of the five stages of grief, before acceptance comes bargaining.

Internally, I'd been trying to use my impact on millions of young

Chinese people as a bargaining chip, as a way to balance out the lack of love shared with family and friends. I had over 50 million followers on Weibo, and I had relentlessly maximized my impact on this group. I even went so far as to build an AI algorithm for discovering and determining what other Weibo messages I should repost, always looking to maximize impact. Yes, I may have skipped out on family time to make public speeches, but think of all the people I had reached. I'd influenced millions of young students and tried to help a once-great country pull itself out of poverty. If you added it all up, wouldn't you say that the good outweighed the bad? Couldn't the gifts I'd given to so many strangers through my work make up for the dearth of love I had shared with those closest to me? Didn't the equation balance out in the end?

Now Master Hsing Yun was kicking the proverbial last leg of the stool out from under me. I tried to explain myself and cast my actions in the best light, based on what they had achieved. But he wasn't interested in the results that my personal well-designed algorithm spat out. He patiently peeled away my layers of excuses and obfuscation. He continually directed the conversation inward, asking me to confront myself with unflinching honesty.

"Kai-Fu, humans aren't meant to think this way. This constant calculating, this quantification of everything, it eats away at what's really inside of us and what exists between us. It suffocates the one thing that gives us true life: love."

"I'm just starting to understand that, Master Hsing Yun," I said, lowering my head, staring at the floor between my two feet.

"Many people understand it," he continued, "but it's much harder to live it. For that we must humble ourselves. We have to feel in our bones just how small we are, and we must recognize that there's nothing greater or more valuable in this world than a simple act of sharing love with others. If we start from there, the rest will begin to fall into place. It's the only way that we can truly become ourselves."

With that, he said goodbye and turned his wheelchair around. I was left with his words echoing in my mind and sinking into my skin. The time since my diagnosis had been a whirlwind of pain, regret, revelation, and doubt. I had come to understand how personally destructive my old ways of thinking had been, and I struggled to replace them with a new way of being human in the world that didn't mimic some aspect of that algorithmic thinking.

In the presence of Master Hsing Yun, I had felt something new. It wasn't so much the answer to a riddle or the solution to a problem. Instead, it was a disposition, a way of understanding oneself and encountering the world that didn't boil down to inputs, outputs, and optimizations.

During my time as a researcher, I had stood on the absolute frontier of human knowledge about artificial intelligence, but I had never been further from a genuine understanding of other human beings or myself. That kind of understanding couldn't be coaxed out of a cleverly constructed algorithm. Rather, it required an unflinching look into the mirror of death and an embrace of that which separated me from the machines that I built: the possibility of love.

SECOND OPINIONS AND SECOND CHANCES

While I wrestled with these stark realizations, the treatment for my cancer proceeded. My first doctor classified the disease as stage IV, the cancer's most advanced stage. On average, patients with fourth-stage lymphoma of my type have around a 50 percent shot of surviving the next five years. I wanted to get a second opinion before beginning treatment, and a friend of mine arranged for me to consult his family doctor, the top hematology practitioner in Taiwan.

It would be a week before I could see that doctor, and in the meantime I continued to conduct my own research on the disease. In my emotional life, I was turning away from the relentless pursuit of quantification and optimization. But as a trained scientist whose life hung in the balance, I couldn't help trying to better understand the disease and quantify my chances of survival. Scouring the internet, I devoured all the information I could find about lymphoma: possible causes, cutting-edge treatments, and long-term survival rates. Through my reading, I came to understand how doctors classify the various stages of lymphoma.

Medical textbooks use the concept of "stages" to describe how advanced cancerous tumors are, with later stages generally corresponding to lower survival rates. In lymphoma, the stage has tradi-

tionally been assigned on the basis of a few straightforward characteristics: Has the cancer affected more than one lymph node? Are
the cancerous lymph nodes both above and below the diaphragm
(the bottom of the rib cage)? Is the cancer found in organs outside
the lymphatic system or in the patient's bone marrow? Traditionally, each answer of "yes" to one of the above questions bumps the
diagnosis up a stage. The fact that my lymphoma had affected over
twenty sites, had spread above and below my diaphragm, and had
entered an organ outside the lymphatic system meant that I was automatically categorized as a stage IV patient.

But what I didn't know at the time of diagnosis was that this
crude method of staging has more to do with what medical students
can memorize than what modern medicine can cure.

Ranking stages based on such simple characteristics of a complex disease is a classic example of the human need to base decisions
on "strong features." Humans are extremely limited in their ability to
discern correlations between variables, so we look for guidance in
a handful of the most obvious signifiers. In making bank loans, for
example, these "strong features" include the borrower's income, the
value of the home, and the credit score. In lymphoma staging, they
simply include the number and location of the tumors.

These so-called strong features really don't represent the most
accurate tools for making a nuanced prognosis, but they're simple
enough for a medical system in which knowledge must be passed
down, stored, and retrieved in the brains of human doctors. Medical research has since identified dozens of other characteristics of
lymphoma cases that make for better predictors of five-year survival
in patients. But memorizing the complex correlations and precise
probabilities of all these predictors is more than even the best medical students can handle. As a result, most doctors don't usually incorporate these other predictors into their own staging decisions.

In the depths of my own research, I found a research paper that
did quantify the predictive power of these alternate metrics. The paper is from a team of researchers at the University of Modena and
Reggio Emilia in Italy, and it analyzed fifteen different variables,
identifying the five features that, considered together, most strongly
correlated to five-year survival. These features included some tradi-

tional measures (such as bone marrow involvement) but also less intuitive measures (are any tumors over 6 cm in diameter? Are hemoglobin levels below 12 grams per deciliter? Is the patient over 60?). The paper then provides average survival rates based on how many of those features a patient exhibited.

To someone trained in artificial intelligence — where even simple algorithms base decisions on hundreds if not thousands of distinct features — this new decision rubric still seemed far from rigorous. It sought to boil down a complex system to just a few features that humans could process. But it also showed that the standard staging metrics were very poor predictors of outcomes and had been created largely to give medical students something they could easily memorize and regurgitate on their tests. The new rubric was far more data-driven, and I leaped at the chance to quantify my own illness by it.

Rifling through stacks of medical reports and test results from the hospital, I dug out the information for each metric: my age, diameter of largest involved node, bone-marrow involvement, β_2-microglobulin status, and hemoglobin levels. Of the five features most strongly correlated to early death, it seemed to appear that I exhibited only one. My eyes frantically scanned the page, sifting through charts and tracing lines between my risk factors and survival rate.

And there it was: while the stage IV diagnosis from the hospital meant a five-year survival rate of just 50 percent, the more detailed and scientific rubric of the research paper bumped that number up to 89 percent.

I kept going back to check and double-check the numbers, and with each confirmation I grew more ecstatic. Nothing inside my body had changed, but I felt that I had been pulled back from the abyss. Later that week, I would visit the top lymphoma expert in Taiwan. He would confirm what the study had indicated: that the designation of my lymphoma as stage IV was misleading, and my illness remained highly treatable. Nothing was certain — I knew that now more than ever — but there was a good chance I would get through this alive. I felt reborn.

There's a certain sensation most people experience right after narrowly avoiding disaster. It's that tingling feeling that crawls over your skin and across your scalp a few seconds after your car skids to a halt on the highway, just a few feet away from an accident. As the adrenaline dissipates and muscles relax, most of us make a silent pledge to never again do whatever it was that we were just doing. It's a pledge we might keep for a couple of days or even weeks before slipping back into old habits.

As I underwent chemotherapy and my cancer went into remission, I too vowed to hold onto the revelations that cancer had given to me. Lying awake at night in the weeks after my diagnosis, I ran over my life again and again, wondering how I had been so blind. I told myself that however much time I had left, I wouldn't let myself be an automaton. I wouldn't live by internal algorithms or seek to optimize variables. I would try to share love with those who had given so much of it to me, not because it achieved a certain goal but just because it felt good and true. I wouldn't seek to be a productivity machine. A loving human being would be enough.

The love of my family during this time served as a constant reminder of this promise and an abiding source of strength during my cancer treatment. Despite years of giving them too little of my own time, when I fell ill my wife, sisters, and daughters all sprang into action to care for me. Shen-Ling was always by my side throughout the exhausting and seemingly endless chemotherapy sessions, tending to my every need and stealing a few hours of sleep leaning against my bedside. Chemotherapy can disrupt digestion, with normal smells and flavors causing nausea or vomiting. When my sisters brought me food, they took careful note of my reaction to each smell or taste, constantly adjusting recipes and tweaking ingredients so that I could enjoy their home-cooked food during treatment. Their selfless love and constant care during this time simply overwhelmed me. It took all the ideas that I had come to understand and turned them into emotions that washed over me and came to live within me.

Since my recovery, I've come to cherish time with those closest to me. Before, when my two daughters came home from college, I would take just a couple of days off work to be with them. Now when they visit from their busy jobs, I take a couple of weeks. Whether on business trips or vacations, I travel with my wife. I spend more time at home taking care of my mother and try to keep my weekends free to see old friends.

I've apologized and tried to mend friendships with those that I have hurt or neglected in the past. I meet with many young people who reach out to me, no longer communicating only through impersonal blasts across my social media accounts. I try to avoid prioritizing these meetings by who "shows potential," doing my best to engage with all people equally, regardless of their status or talents.

I no longer think about what will be written on my tombstone. That's not because I avoid thinking about death. I'm now more aware than ever that we all live in direct and constant relationship to our own mortality. It's because I know that my tombstone is just a piece of stone, a lifeless rock that can't compare with the people and memories that make up the rich tapestry of a human life. I recognize that I'm just beginning to learn what so many people around me understood intuitively all their lives. But simple as these realizations are, they have transformed my life.

They've also transformed how I view the relationship between people and machines, between human hearts and artificial minds. This transformation crept up on me as I reflected on the process of my illness: the PET scan, the diagnosis, my own anguish, and the physical and emotional healing that followed. I've come to realize that my cure came in two parts, one technological and one emotional, each of which will form a pillar of our AI future as I explain in the next chapter.

I have great respect and deep appreciation for the medical professionals who led my treatment. They put years of experience and cutting-edge medical technology to the task of beating back the lymphoma that grew within me. Their knowledge of this illness and their ability to craft a personalized treatment regimen likely saved my life.

And yet, that was only half of the cure for what ailed me. I wouldn't be here today if it weren't for medical technology and the data-driven practitioners who use it to save lives. But I wouldn't be sharing this story with you if it weren't for Shen-Ling, my sisters, and my own mother, who through quiet example showed me what it means to lead a life of selflessly sharing love.

Or people like Bronnie Ware, whose heartfelt book on the regrets of the dying gave me life at my weakest moment. Or Master Hsing Yun, whose wisdom shook me from my career delusions and forced me to truly confront my own ego. Without these unquantifiable, nonoptimizable connections to other people, I would never have learned what it truly means to be human. Without them, I would never have reordered my priorities and reoriented my own life. I soon began working less and spending more time with the people in my life. I stopped trying to quantify the impact of each action — who I took meetings with, who I wrote back to, who I spent time with — and instead aimed to treat all those around me equally. This shift in the way I treated others wasn't just beneficial to them; it filled me with a sense of wholeness, satisfaction, and calm that the hollow accomplishments of my career never could.

The reality is that it will not be long until AI algorithms can perform many of the diagnostic functions of medical professionals. Those algorithms will pinpoint illness and prescribe treatments more effectively than any single human can. In some cases, doctors will use these equations as a tool. In some cases, the algorithms may replace the doctor entirely.

But the truth is, there exists no algorithm that could replace the role of my family in my healing process. What they shared with me is far simpler — and yet so much more profound — than anything AI will ever produce.

For all of AI's astounding capabilities, the one thing that only humans can provide turns out to also be exactly what is most needed in our lives: love. It's that moment when we see our newborn babies, the feeling of love at first sight, the warm feeling from friends who listen to us empathetically, or the feeling of self-actualization when we help someone in need. We are far from understanding the human heart, let alone replicating it. But we do know that humans are uniquely able to love and be loved, that humans want to love and be loved, and that loving and being loved are what makes our lives worthwhile.

This is the synthesis on which I believe we must build our shared future: on AI's ability to think but coupled with human beings' ability to love. If we can create this synergy, it will let us harness the undeniable power of artificial intelligence to generate prosperity while also embracing our essential humanity.

This isn't something that will come naturally. Building this future for ourselves — as people, countries, and a global community — will require that we reimagine and reorganize our societies from the ground up. It will take social unity, creative policies, and human empathy, but if achieved, it could turn a moment of outright crisis into an unparalleled opportunity.

Never has the potential for human flourishing been higher — or the stakes of failure greater.

8

*

A BLUEPRINT FOR HUMAN COEXISTENCE WITH AI

While I was undergoing chemotherapy for my cancer in Taiwan, an old friend of mine who is a serial entrepreneur came to me with a problem at his latest startup. He had already founded and sold off several successful consumer technology companies, but as he grew older he wanted to do something more meaningful, that is, he wanted to build a product that would serve the people that technology startups had often ignored. Both my friend and I were entering the age at which our parents needed more help going about their daily lives, and he decided to design a product that would make life easier for the elderly.

What he came up with was a large touchscreen mounted on a stand that could be placed next to an elderly person's bed. On the screen were a few simple and practical apps connected to services that elderly people could use: ordering food delivery, playing their favorite soap operas on the TV, calling their doctor, and more. Older people often struggle to navigate the complexities of the internet or to manipulate the small buttons of a smartphone, so my friend made everything as simple as possible. All the apps required just a couple of clicks, and he even included a button that let the elderly users directly call up a customer-service agent to guide them through using their device.

It sounded like a wonderful product, one that would have a real market right now. Sadly, there are many adult children in China and elsewhere who are too busy with work to devote time to taking care of their aging parents. They may experience a sense of guilt about the importance of filial piety, but when it comes down to it, they just don't feel they can find the time to care for their parents in an adequate way. The touchscreen would make for a nice substitute.

But after deploying a trial version of his product, my friend discovered he had a problem. Of all the functions available on the device, the one that received by far the most use wasn't the food delivery, TV controls, or doctor's consultation. It was the customer-service button. The company's customer-service representatives found themselves overwhelmed by a flood of incoming calls from the elderly. What was going on here? My friend had made the device as simple as possible to use — were his users still unable to navigate the one-click process onscreen?

Not at all. After consulting with the customer-service representatives, he found that people weren't calling in because they couldn't navigate the device. They were calling simply because they were lonely and wanted someone to talk to. Many of the elderly users had children who worked to ensure that all of their material needs were met: meals were delivered, doctors' appointments were arranged, and prescriptions were picked up. But once those material needs were taken care of, what these people wanted more than anything was true human contact, another person to trade stories with and relate to.

My friend relayed this "problem" to me just as I was waking up to my own realizations about the centrality of love to the human experience. If he had come to me just a few years earlier, I likely would have recommended some technical fix, maybe something like an AI chat bot that could simulate a basic conversation well enough to fool the human on the other end. But as I recovered from my illness and awakened to the looming AI crises of jobs and meaning, I began to see things differently.

In that touchscreen device and that unmet desire for human contact, I saw the first sketches of a blueprint for coexistence between people and artificial intelligence. Yes, intelligent machines will increasingly be able to do our jobs and meet our material needs, disrupting industries and displacing workers in the process. But there remains one thing that only human beings are able to create and share with one another: love.

With all of the advances in machine learning, the truth remains that we are still nowhere near creating AI machines that feel any emotions at all. Can you imagine the elation that comes from beating a world champion at the game you've devoted your whole life to mastering? AlphaGo did just that, but it took no pleasure in its success, felt no happiness from winning, and had no desire to hug a loved one after its victory. Despite what science-fiction films like *Her* — in which a man and his artificially intelligent computer operating system fall in love — portray, AI has no ability or desire to love or be loved. The actress Scarlett Johansson may have been able to convince you otherwise in that film, but only because she is a human being who drew on her experience of love to create and communicate those feelings to you.

Imagine a situation in which you informed a smart machine that you were going to pull its plug, and then changed your mind and gave it a reprieve. The machine would not change its outlook on life or vow to spend more time with its fellow machines. It would not grow emotionally or discover the value in loving and serving others.

It is in this uniquely human potential for growth, compassion, and love where I see hope. I firmly believe we must forge a new synergy between artificial intelligence and the human heart, and look for ways to use the forthcoming material abundance generated by artificial intelligence to foster love and compassion in our societies.

If we can do these things, I believe there is a path toward a future of both economic prosperity and spiritual flourishing. Navigating that path will be tricky, but if we are able to unite behind this common goal, I believe humans will not just survive in the age of AI. We will thrive like never before.

A TRIAL BY FIRE AND THE NEW SOCIAL CONTRACT

The challenges before us remain immense. As I outlined in chapter 6, within fifteen years I predict that we will technically be able to automate 40 to 50 percent of all jobs in the United States. That does not mean all of those jobs will disappear overnight, but if the markets are left to their own devices, we will begin to see massive pressure on working people. China and other developing countries may differ slightly in the timing of those impacts, lagging or leading in job losses depending on the structures of their economies. But the overarching trend remains the same: rising unemployment and widening inequality.

Techno-optimists will point to history, citing the Industrial Revolution and the nineteenth-century textile industry as "proof" that things always work out for the best. But as we've seen, this argument stands on increasingly shaky ground. The coming scale, pace, and skill-bias of the AI revolution mean that we face a new and historically unique challenge. Even if the most dire predictions of unemployment do not materialize, AI will take the growing wealth inequality of the internet age and accelerate it tremendously.

We are already witnessing the way that stagnant wages and growing inequality can lead to political instability and even violence. As AI rolls out across our economies and societies, we risk aggravating and quickening these trends. Labor markets have a way of balancing themselves out in the long run, but getting to that promised long run requires we first pass through a trial by fire of job losses and growing inequality that threaten to derail the process.

Meeting these challenges means we cannot afford to passively react. We must proactively seize the opportunity that the material wealth of AI will grant us and use it to reconstruct our economies and rewrite our social contracts. The epiphanies that emerged from my experience with cancer were deeply personal, but I believe they also gave me a new clarity and vision for how we can approach these problems together.

Building societies that thrive in the age of AI will require substantial changes to our economy but also a shift in culture and values. Centuries of living within the industrial economy have conditioned many of us to believe that our primary role in society (and even our identity) is found in productive, wage-earning work. Take that away and you have broken one of the strongest bonds between a person and his or her community. As we transition from the industrial age to the AI age, we will need to move away from a mindset that equates work with life or treats humans as variables in a grand productivity optimization algorithm. Instead, we must move toward a new culture that values human love, service, and compassion more than ever before.

No economic or social policy can "brute force" a change in our hearts. But in choosing different policies, we can reward different behaviors and start to nudge our culture in different directions. We can choose a purely technocratic approach — one that sees each of us as a set of financial and material needs to be satisfied — and simply transfer enough cash to all people so that they don't starve or go homeless. In fact, this notion of universal basic income seems to be becoming more and more popular these days.

But in making that choice I believe we would both devalue our own humanity and miss out on an unparalleled opportunity. Instead, I want to lay out proposals for how we can use the economic bounty created by AI to double-down on what makes us human. Doing this will require rewriting our fundamental social contracts and restructuring economic incentives to reward socially productive activities in the same way that the industrial economy rewarded economically productive activities.

This won't be easy. It will need a multifaceted, all-hands-on-deck approach to economic and social transformation. That approach will rely on input from all corners of society and must be based on constant exploration and bold experimentation. Even with our best efforts, there remains no guarantee of a smooth transition. But both the cost of failure and the potential rewards of success are too great not to try.

Let's begin that process.

First, I want to examine three of the most popular policy suggestions for adapting to the AI economy, many of them emanating from Silicon Valley. These three are largely "technical fixes," tweaks to policy and business models that seek to smooth the transition but do not actually shift the culture. After examining the uses and weaknesses of these technical fixes, I propose three analogous changes that I believe will both alleviate the jobs issues while also pushing us toward a deeper social evolution.

Instead of just implementing mere technical fixes, these constitute new approaches to job creation within the private sector, affecting investing and government policy. These approaches take as their goal not just keeping humans one step ahead of AI automation but actually opening new avenues to increased prosperity and human flourishing. Together, I believe they lay the groundwork for a new social contract that uses AI to build a more humanistic world.

THE CHINESE PERSPECTIVE ON AI AND JOBS

Before diving into the technical fixes proposed by Silicon Valley, let's first look at how this conversation is unfolding in China. To date, China's tech elite have said very little about the possible negative impact of AI on jobs. Personally, I don't believe this silence is due to any desire to hide the dark truth from the masses — I think they genuinely believe there is nothing to fear in the jobs impact of AI advances. In this sense, China's tech elites are aligned with the techno-optimistic American economists who believe that in the long run, technology always leads to more jobs and greater prosperity for all.

Why does a Chinese entrepreneur believe in that with such conviction? For the past forty years, Chinese people have watched as their country's technological progress acted as the rising tide that lifted all boats. The Chinese government has long emphasized technological advances as key to China's economic development, and that model has proved highly successful in recent decades, moving China from a predominantly agricultural society to an industrial juggernaut and now an innovation powerhouse. Inequality has certainly increased over this same period of time, but those downsides have paled in comparison to the broad-based improvement in livelihoods. It makes a stark contrast to the stagnation and decline felt in many segments of American society, part of the "great decoupling" between productivity and wages we explored in previous chapters. It also helps explain why Chinese technologists appear unconcerned with the potential jobs impact of their innovations.

Even among the Chinese entrepreneurs who do foresee a negative AI impact, there is a pervasive sense that the Chinese government will take care of all the displaced workers. This idea isn't without basis. During the 1990s, China undertook a series of wrenching reforms to its bloated state-owned companies, shedding millions of workers from government payrolls. But despite the massive labor-market disruptions, the strength of the national economy and a far-reaching government effort to help workers manage the transition combined to successfully transform the economy without widespread unemployment. Looking into the AI future, many technologists and policymakers share an unspoken belief that these same mechanisms will help China avoid an AI-induced job crisis.

Personally, I believe these predictions are too optimistic, so I am working to raise consciousness in China, as I am in the United States, regarding the momentous employment challenges that await us in the age of AI. It is important that Chinese entrepreneurs, technologists, and policymakers take these challenges seriously and begin laying the groundwork for creative solutions. But the cultural mentality described above — one that is reinforced by four decades of growing prosperity — means that we see little discussion of the crisis in China and even less in the way of proposed solutions. To engage with that conversation, we must turn again to Silicon Valley.

THE THREE R'S: REDUCE, RETRAIN, AND REDISTRIBUTE

Many of the proposed technical solutions for AI-induced job losses coming out of Silicon Valley fall into three buckets: retraining workers, reducing work hours, or redistributing income. Each of these approaches aims to augment a different variable within the labor markets (skills, time, compensation) and also embodies different assumption about the speed and severity of job losses.

Those advocating the *retraining* of workers tend to believe that AI will slowly shift what skills are in demand, but if workers can adapt their abilities and training, then there will be no decrease in the need for labor. Those advocates of *reducing work hours* believe that AI will reduce the demand for human labor and feel that this impact could be absorbed by moving to a three- or four-day work week, spreading the jobs that do remain over more workers. The *redistribution* camp tends to be the most dire in their predictions of AI-induced job losses. Many of them predict that as AI advances, it will so thoroughly displace or dislodge workers that no amount of training or tweaking hours will be sufficient. Instead, we will have to adopt more radical redistribution schemes to support unemployed workers and spread the wealth created by AI. Next, I will take a closer look at the value and pitfalls of each of these approaches.

Advocates of job retraining often point to two related trends as crucial for creating an AI-ready workforce: online education and "lifelong learning." They believe that with the proliferation of online education platforms — both free and paid — displaced workers will have unprecedented access to training materials and instruction for new jobs. These platforms — video streaming sites, online coding academies, and so on — will give workers the tools they need to become lifelong learners, constantly updating their skills and moving into new professions that are not yet subject to automation. In this envisioned world of fluid retraining, unemployed insurance brokers can use online education platforms like Coursera to become software programmers. And when that job becomes automated, they can use those same tools to retrain for a new position that remains out of reach for AI, perhaps as an algorithm engineer or as a psychologist.

Lifelong learning via online platforms is a nice idea, and I believe retraining workers will be an important piece of the puzzle. It can particularly help those individuals within the bottom-right quadrant of our risk-of-replacement charts from chapter 6 (the "Slow Creep" zone) stay ahead of AI's ability to think creatively or work in unstructured environments. I also like that this method can give these workers a sense of personal accomplishment and agency in their own lives.

But given the depth and breadth of AI's impact on jobs, I fear this approach will be far from enough to solve the problem. As AI steadily conquers new professions, workers will be forced to change occupations every few years, rapidly trying to acquire skills that it took others an entire lifetime to build up. Uncertainty over the pace and path of automation makes things even more difficult. Even AI experts have difficulty predicting exactly which jobs will be subject to automation in the coming years. Can we really expect a typical worker choosing a retraining program to accurately predict which jobs will be safe a few years from now?

I fear workers will find themselves in a state of constant retreat, like animals fleeing relentlessly rising flood waters, anxiously hopping from one rock to another in search of higher ground. Retraining will help many people find their place in the AI economy, and we must experiment with ways to scale this up and make it widely available. But I believe we cannot count on this haphazard approach to address the macro-level disruptions that will sweep over labor markets.

To be clear, I do believe that education is the best long-term solution to the AI-related employment problems we will face. The previous millennia of progress have demonstrated human beings' incredible ability both to innovate technically and to adapt to those innovations by training ourselves for new kinds of work. But the scale and speed of the coming changes from AI will not give us the luxury of simply relying on educational improvements to help us keep pace with the changing demands of our own inventions.

Recognition of the scale of these disruptions has led people like Google cofounder Larry Page to advocate a more radical proposition: let's move to a four-day work week or have multiple people "share" the same job. In one version of this proposal, a single full-time job could be split into several part-time jobs, sharing the increasingly scarce resource of jobs across a larger pool of workers. These approaches would likely mean reduced take-home pay for most workers, but these changes could at least help people avoid outright unemployment.

Some creative approaches to work-sharing have already been implemented. Following the 2008 financial crisis, several U.S. states implemented work-sharing arrangements to avoid mass layoffs at companies whose business suddenly dried up. Instead of laying off a portion of workers, companies reduced hours for several workers by 20 to 40 percent. The local government then compensated those workers for a certain percentage of their lost wages, often 50 percent. This approach worked well in some places, saving employees and companies the disruptions of firing and rehiring at the whim of the business cycle. It also potentially saved local governments money that would have gone to paying full unemployment benefits.

Work-share arrangements could blunt job losses, particularly for professions in the "Human Veneer" quadrant of our risk-of-replacement graphs, where AI performs the main job task but only a smaller number of workers are needed to interface with customers. If executed well, these arrangements could act as government subsidies or incentives to keep more workers on the company payroll.

But while this approach works well for short-term disruptions, it may lose traction in the face of AI's persistent and nonstop decimation of jobs. Existing work-share programs only supplement a portion of lost wages, meaning workers still saw a net decline in income. Workers may accept this knock to their income during a temporary economic crisis, but no one desires stagnation or downward mobility over the long term. Telling a worker making $20,000 a year that they can now work four days a week and earn $16,000 is really a nonstarter. More creative versions of these programs could correct for this, and I encourage companies and governments to continue experimenting with them. But I fear this kind of approach will be far from sufficient to address the long-term pressures that AI will bring to the labor market. For that, we may have to adopt more radical redistributive measures.

THE BASICS OF UNIVERSAL BASIC INCOME

Currently, the most popular of these methods of redistribution is, as mentioned earlier, the universal basic income (UBI). At its core, the idea is simple: every citizen (or every adult) in a country receives a regular income stipend from the government — no strings attached. A UBI would differ from traditional welfare or unemployment benefits in that it would be given to everyone and would not be subject to time limits, job-search requirements, or any constraints in how it could be spent. An alternate proposal, often called a guaranteed minimum income (GMI), calls for giving the stipend only to the poor, turning it into an "income floor" below which no one could fall but without the universality of a UBI.

Funding for these programs would come from steep taxes on the winners of the AI revolution: major technology companies; legacy corporations that adapted to leverage AI; and the millionaires, billionaires, and perhaps even trillionaires who cashed in on these companies' success. The size of the stipend given is a matter of debate among proponents. Some people argue for keeping it very small — perhaps just $10,000 per year — so that workers still have a strong incentive to find a real job. Others view the stipend as a full replacement for the lost income of a regular job. In this view, a UBI could become a crucial step toward creating a "leisure society," one in which people are fully liberated from the need to work, and free to pursue their own passions in life.

Discussion of a UBI or GMI in the United States dates back to the 1960s, when it won support from people as varied as Martin Luther King Jr. and Richard Nixon. At the time, advocates saw a GMI as a simple way to end poverty, and in 1970 President Nixon actually came close to passing a bill that would have granted each family enough money to raise itself above the poverty line. But following Nixon's unsuccessful push, discussion of a UBI or GMI largely dropped out of public discourse.

That is, until Silicon Valley got excited about it. Recently, the idea has captured the imagination of the Silicon Valley elite, with giants of the industry like the prestigious Silicon Valley startup accelerator Y Combinator president Sam Altman and Facebook cofounder Chris Hughes sponsoring research and funding basic income pilot programs. Whereas GMI was initially crafted as a cure for poverty in normal economic times, Silicon Valley's surging interest in the programs sees them as solutions for widespread technological unemployment due to AI.

The bleak predictions of broad unemployment and unrest have put many of the Silicon Valley elite on edge. People who have spent their careers preaching the gospel of disruption appear to have suddenly woken up to the fact that when you disrupt an industry, you also disrupt and displace real human beings within it. Having founded and funded transformative internet companies that also contributed to gaping inequality, this cadre of millionaires and billionaires appear determined to soften the blow in the age of AI.

To these proponents, massive redistribution schemes are potentially all that stand between an AI-driven economy and widespread joblessness and destitution. Job retraining and clever scheduling are hopeless in the face of widespread automation, they argue. Only a guaranteed income will let us avert disaster during the jobs crisis that looms ahead.

How exactly a UBI would be implemented remains to be seen. A research organization associated with Y Combinator is currently running one pilot program in Oakland, California, that gives a thousand families a stipend of a thousand dollars each month for three to five years. The research group will track the well-being and activities of those families through regular questionnaires, comparing them with a control group that receives just fifty dollars per month.

Many in Silicon Valley see the program through the lens of their own experience as entrepreneurs. They envision the money not only as a kind of broad safety net but as an "investment in the startup of you," or as one tech writer put it, "VC for the people." In this worldview, a UBI would give unemployed people a little "personal angel investment" with which they could start a new business or learn a new skill. In his 2017 Harvard commencement speech, Mark Zuckerberg aligned himself with this vision of UBI, arguing that we should explore a UBI so that "everyone has a cushion to try new ideas."

From my perspective, I can understand why the Silicon Valley elite have become so enamored with the idea of a UBI: it is a simple, technical solution to an enormous and complex social problem of their own making. But adopting a UBI would constitute a major change in our social contract, one that we should think through very carefully and most critically. While I support certain guarantees that basic needs will be met, I also believe embracing a UBI as a cure-all for the crisis we face is a mistake and a massive missed opportunity. To understand why, we must truly look at the motivations for the frenzy of interest in UBI and also think hard about what kind of a society it may create.

SILICON VALLEY'S "MAGIC WAND" MENTALITY

In observing Silicon Valley's surge in interest around UBI, I believe some of that advocacy has emerged from a place of true and genuine concern for those who will be displaced by new technologies. But I worry that there's also a more self-interested component: Silicon Valley entrepreneurs know that their billions in riches and their role in instigating these disruptions make them an obvious target of mob anger if things ever spin out of control. With that fear fresh in their minds, I wonder if this group has begun casting about for a quick fix to problems ahead.

The mixed motivations of these people shouldn't lead us to outright dismiss the solutions they put forth. This group, after all, includes some of the most creative business and engineering minds in the world today. Silicon Valley's tendency to dream big, experiment, and iterate will all be helpful as we navigate these uncharted waters.

But an awareness of these motivations should sharpen our critical engagement with proposals like UBI. We should be aware of the cultural biases that engineers and investors bring with them when tackling a new problem, particularly one with profound social and human dimensions. Most of all, when evaluating these proposed solutions, we must ask what exactly they're trying to achieve. Are they seeking to ensure that this technology genuinely and truly benefits all people across society? Or are they looking only to avert a worst-case scenario of social upheaval? Are they willing to put in the legwork needed to build new institutions or merely looking for a quick fix that will assuage their own consciences and absolve them of responsibility for the deeper psychological impacts of automation?

I fear that many of those in Silicon Valley are firmly in the latter camp. They see UBI as a "magic wand" that can make disappear the myriad economic, social, and psychological downsides of their exploits in the AI age. UBI is the epitome of the "light" approach to problem-solving so popular in the valley: stick to the purely digital sphere and avoid the messy details of taking action in the real world. It tends to envision that all problems can be solved through a tweaking of incentives or a shuffling of money between digital bank accounts.

Best of all, it doesn't place any further burden on researchers to think critically about the societal impacts of the technologies they build; as long as everyone gets that monthly dose of UBI, all is well. The tech elite can go on doing exactly what they planned to do in the first place: building innovative companies and reaping massive financial rewards. Sure, higher taxes required to fund a UBI will cut into those profits to a certain degree, but the vast majority of the financial benefits from AI will still accrue to this elite group.

Seen in this manner, UBI isn't a constructive solution that leverages AI to build a better world. It's a painkiller, something to numb and sedate the people who have been hurt by the adoption of AI. And that numbing effect goes both ways: not only does it ease the pain for those displaced by technology; it also assuages the conscience of those who do the displacing.

As I've said before, some form of guaranteed income may be necessary to put an economic floor under everyone in society. But if we allow this to be the endgame, we miss out on the great opportunity presented to us by this technology. Instead of simply falling back on a painkiller like a UBI, we must proactively seek and find ways of utilizing AI to double-down on that which separates us from machines: love.

Admittedly, this won't be easy. It will require creative and different approaches. Executing on these approaches will take a lot of legwork and "heavy" solutions, reaching beyond the digital sphere and into the not-so-neat details of the real world. But if we commit to doing the hard work now, I believe we have a shot at not just avoiding disaster but of cultivating the same humanistic values that I rediscovered during my own encounter with mortality.

MARKET SYMBIOSIS: OPTIMIZATION
TASKS AND HUMAN TOUCH

The private sector is leading the AI revolution, and, in my mind, it must also take the lead in creating the new, more humanistic jobs that power it. Some of these will emerge through the natural functioning of the free market, while others will require conscious efforts by those motivated to make a difference.

Many of the jobs created by the free market will grow out of a natural symbiosis between humans and machines. While AI handles the routine optimization tasks, human beings will bring the personal, creative, and compassionate touch. This will involve the redefinition of existing occupations or the creation of entirely new professions in which people team up with machines to deliver services that are both highly efficient and eminently human. In the risk-of-replacement graphs from chapter 6, we expect to see the upper-left quadrant ("Human Veneer") offer the greatest opportunity for human-AI symbiosis: AI will do the analytical thinking, while humans will wrap that analysis in warmth and compassion. In that same chart, the two quadrants on the right-hand side of the graph ("Slow Creep" and "Safe Zone") also provide opportunities for AI tools to enhance creativity or decision-making, though over time, the two left-side AI-centric circles will grow toward the right as AI improves.

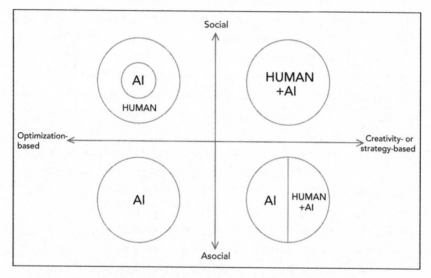

Human–AI coexistence in the labor market

A clear example of human-AI symbiosis for the upper-left-hand quadrant can be found in the field of medicine. I have little doubt that AI algorithms will eventually far surpass human doctors in their ability to diagnose disease and recommend treatments. Legacy institutions — medical schools, professional associations, and hospitals — may slow down the adoption of these diagnostic tools, using them only in narrow fields or strictly as reference tools. But in a matter of a few decades, I'm confident that the accuracy and efficiency gains will be so great that AI-driven diagnoses will take over eventually.

One response to this would be to get rid of doctors entirely, replacing them with machines that take in symptoms and spit out diagnoses. But patients don't want to be treated by a machine, a black box of medical knowledge that delivers a cold pronouncement: "You have fourth-stage lymphoma and a 70 percent likelihood of dying within five years." Instead, patients will desire — and I believe the market will create — a more humanistic approach to medicine.

Traditional doctors could instead evolve into a new profession, one that I'll call a "compassionate caregiver." These medical professionals would combine the skills of a nurse, medical technician, social worker, and even psychologist. Compassionate caregivers would be trained not just in operating and understanding the diagnostic tools but also in communicating with patients, consoling them in times of trauma, and emotionally supporting them throughout their treatment. Instead of simply informing patients of their objectively optimized chances of survival, they could share encouraging stories, saying "Kai-Fu had the same lymphoma as you and he survived, so I believe you can too."

These compassionate caregivers wouldn't compete with machines in their ability to memorize facts or optimize treatment regimens. In the long run, that's a losing battle. Compassionate caregivers would be well trained, but in activities requiring more emotional intelligence, not as mere vessels for the canon of medical knowledge. They would form a perfect complement to the machine, giving patients unparalleled accuracy in their diagnoses as well as the human touch that is so often missing from our hospitals today. In this human-machine symbiosis created by the free market, we would inch our society ahead in a direction of being a little kinder and a little more loving.

Best of all, the emergence of compassionate caregivers would dramatically increase both the number of jobs and the total amount of medical care given. Today, the scarcity of trained doctors drives up the cost of healthcare and drives down the amount of quality care delivered around the world. Under current conditions of supply and demand, it's simply not cost-feasible to increase the number of doctors. As a result, we strictly ration the care they deliver. No one wants to go wait in line for hours just to have a few minutes with a doctor, meaning that most people only go to hospitals when they feel it's absolutely necessary. While compassionate caregivers will be well-trained, they can be drawn from a larger pool of workers than doctors and won't need to undergo the years of rote memorization that is required of doctors today. As a result, society will be able to cost-effectively support far more compassionate caregivers than there are doctors, and we would receive far more and better care.

Similar synergies will emerge in many other fields: teaching, law, event planning, and high-end retail. Paralegals at law firms could hand their routine research tasks off to algorithms and instead focus on communicating more with clients and making them feel cared for. AI-powered supermarkets like the Amazon Go store may not need cashiers anymore, so they could greatly upgrade the customer experience by hiring friendly concierges like the one I described in chapter 5.

For those in professional sectors, it will be imperative that they adopt and learn to leverage AI tools as they arrive. As with any technological revolution, many workers will find the new tools both imperfect in their uses and potentially threatening in their implications. But these tools will only improve with time, and those who seek to compete against AI on its own terms will lose out. In the long run, resistance may be futile, but symbiosis will be rewarded.

Finally, the internet-enabled sharing economy will contribute significantly to alleviating job losses and redefining work for the AI age. We'll see more people step out of traditional careers that are being taken over by algorithms, instead using new platforms that apply the "Uber model" to a variety of services. We see this already in Care.com, an online platform for connecting caregivers and customers, and I believe we will see a blossoming of analogous models in education and other fields. Many mass-market goods and services will be captured by data and optimized by algorithms, but some of the more piecemeal or personalized work within the sharing economy will remain the exclusive domain of humans.

In the past, this type of work was constrained by the bureaucratic costs of running a vertical company that attracted customers, dispatched workers, and kept everyone on the payroll even when there wasn't work to be done. The platformatization of these industries dramatically increases their efficiency, increasing total demand

and take-home pay for the service workers themselves. Adding AI to the equation — as ride-hailing companies like Didi and Uber have already done — will only further boost efficiency and attract more workers.

Beyond the established roles in the sharing economy, I'm confident we will see entirely new service jobs emerge that we can hardly imagine today. Explain to someone in the 1950s what a "life coach" was and they'd probably think you were goofy. Likewise, as AI frees up our time, creative entrepreneurs and ordinary people will leverage these platforms to create new kinds of jobs. Perhaps people will hire "season changers" who redecorate their closets every few months, scenting them with flowers and aromas that match the mood of the season. Or environmentally conscious families will hire "home sustainability consultants" to meet with the family and explore creative and fun ways for the household to reduce its environmental footprint.

But despite all these new possibilities created by profit-seeking businesses, I'm afraid the operations of the free market alone will not be enough to offset the massive job losses and gaping inequality on the horizon. Private companies already create plenty of human-centered service jobs — they just don't pay well. Economic incentives, public policies, and cultural dispositions have meant that many of the most compassion-filled professions existing today often lack job security or basic dignity.

The U.S. Bureau of Labor Statistics has found that home health aides and personal care aides are the two fastest growing professions in the country, with an expected growth of 1.2 million jobs by 2026. But annual income in these professions averages just over $20,000. Other humanistic labors of love — stay-at-home parenting, caring for aging or disabled relatives — aren't even considered a "job" and receive no formal compensation.

These are exactly the kinds of loving and compassionate activities that we should embrace in the AI economy, but the private sector has proven inadequate so far at fostering them. There may come a day when we enjoy such material abundance that economic incentives are no longer needed. But in our present economic and cultural moment, money still talks. Orchestrating a true shift in culture will require not just creating these jobs but turning them into true careers with respectable pay and greater dignity.

Encouraging and rewarding these prosocial activities means going beyond the market symbiosis of the private sector. We will need to reenergize these industries through service sector impact investing and government policies that nudge forward a broader shift in cultural values.

FINK'S LETTER AND THE NEW IMPACT INVESTING

When a man overseeing $5.7 *trillion* speaks, the global business community tends to listen. So when BlackRock founder Larry Fink, head of the world's largest asset management company, posted a letter to CEOs demanding greater attention to social impact, it sent shockwaves through corporations around the globe. In the letter, titled "A Sense of Purpose," Fink wrote,

> We . . . see many governments failing to prepare for the future, on issues ranging from retirement and infrastructure to automation and worker retraining. As a result, society increasingly is turning to the private sector and asking that companies respond to broader societal challenges. . . . Society is demanding that companies, both public and private, serve a social purpose. . . . Companies must benefit all of their stakeholders, including shareholders, employees, customers, and the communities in which they operate.

Fink's letter dropped just days before the 2018 World Economic Forum, an annual gathering of the global financial elite in Davos, Switzerland. I was attending the forum and watched as CEOs anxiously discussed the stern warning from a man whose firm controlled substantial ownership stakes in their companies. Many publicly professed sympathy for Fink's message but privately declared his emphasis on broader social welfare to be anathema to the logic of private enterprise.

Looked at narrowly enough, they're right: publicly traded companies are in it to win it, bound by fiduciary duties to maximize profits. But in the age of AI, this cold logic of dollars and cents simply can't hold. Blindly pursuing profits without any thought to social impact won't just be morally dubious; it will be downright dangerous.

Fink referenced automation and job retraining multiple times in his letter. As an investor with interests spanning the full breadth of the global economy, he sees that dealing with AI-induced displacement is not something that can be left entirely up to free markets. Instead, it is imperative that we reimagine and reinvigorate corporate social responsibility, impact investing, and social entrepreneurship.

In the past, these were the kinds of things that businesspeople merely dabbled in when they had time and money to spare. Sure, they think, why not throw some money into a microfinance startup or buy some corporate carbon offsets so we can put out a happy press release touting it. But in the age of AI, we will need to seriously deepen our commitment to — and broaden our definition of — these activities. Whereas these have previously focused on feel-good philanthropic issues like environmental protection and poverty alleviation, social impact in the age of AI must also take on a new dimension: the creation of large numbers of service jobs for displaced workers.

As a venture-capital investor, I see a particularly strong role for a new kind of impact investing. I foresee a venture ecosystem emerging that views the creation of humanistic service-sector jobs as a good in and of itself. It will steer money into human-focused service projects that can scale up and hire large numbers of people: lactation consultants for postnatal care, trained coaches for youth sports, gatherers of family oral histories, nature guides at national parks, or conversation partners for the elderly. Jobs like these can be meaningful on both a societal and personal level, and many of them have the potential to generate real revenue — just not the 10,000 percent returns that come from investing in a unicorn technology startup.

Kick-starting this ecosystem will require a shift in mentality for VCs who participate. The very idea of venture capital has been built around high risks and exponential returns. When an investor puts money into ten startups, they know full well that nine of them most likely will fail. But if that one success story turns into a billion-dollar company, the exponential returns on that one investment make the fund a huge success. Driving those exponential returns are the unique economics of the internet. Digital products can be scaled up infinitely with near-zero marginal costs, meaning the most successful companies achieve astronomical profits.

Service-focused impact investing, however, will need to be different. It will need to accept *linear returns* when coupled with meaningful job creation. That's because human-driven service jobs simply cannot achieve these exponential returns on investment. When someone builds a great company around human care work, they cannot digitally replicate these services and blast them out across the globe. Instead, the business must be built piece by piece, worker by worker. The truth is, traditional VCs wouldn't bother with these kinds of linear companies, but these companies will be a key pillar in building an AI economy that creates new jobs and fosters human connections.

There will of course be failures, and returns will never match pure technology VC funds. But that should be fine with those involved. The ecosystem will likely be staffed by older VC executives who are looking to make a difference, or possibly by younger VC types who are taking a "sabbatical" or doing "pro bono" work. They will bring along their keen instincts for picking entrepreneurs and building companies, and will put them to work on these linear service companies. The money behind the funds will likely come from governments looking to efficiently generate new jobs, as well as companies doing corporate social responsibility.

Together, these players will create a unique ecosystem that is much more jobs-focused than pure philanthropy, much more impact-focused than pure venture capital. If we can pull together these different strands of socially conscious business, I believe we'll be able to weave a new kind of employment safety net, all while building communities that foster love and compassion.

BIG CHANGES AND BIG GOVERNMENT

And yet, for all the power of the private market and the good intentions of social entrepreneurs, many people will still fall through the cracks. We need look no further than the gaping inequality and destitute poverty in so much of the world today to recognize that markets and moral imperatives are not enough. Orchestrating a fundamental change in economic structures often requires the full force of governmental power. If we hope to write a new social contract for the age of AI, we will need to pull on the levers of public policy.

There are some in Silicon Valley who see this as the point where UBI comes into play. Faced with inadequate job growth, the government must provide a blanket guarantee of economic security, a cash transfer that can save displaced workers from destitution and which will also save the tech elite from having to do anything else about it.

The unconditional nature of the transfer fits with the highly individualistic, live-and-let-live libertarianism that undergirds much of Silicon Valley. Who is the government, UBI proponents ask, to tell people how to spend their time? Just give them the money and let them figure it out on their own. It's an approach that matches how the tech elite tend to view society as a whole. Looking outward from Silicon Valley, they often see the world in terms of "users" rather than citizens, customers rather than members of a community.

I have a different vision. I don't want to live in a society divided into technological castes, where the AI elite live in a cloistered world of almost unimaginable wealth, relying on minimal handouts to keep the unemployed masses sedate in their place. I want to create a system that provides for *all* members of society, but one that also uses the wealth generated by AI to build a society that is more compassionate, loving, and ultimately human.

Achieving this outcome will definitely require creative thinking and complex policymaking, but the inspiration driving that process often comes from unlikely places. For me, it began back at Fo Guang Shan, the monastery in Taiwan that I discussed in the previous chapter.

THE CHAUFFEUR CEO

The morning sun had not yet crept over the horizon as I walked across the monastery's massive grounds to see Master Hsing Yun. It was the morning on which I'd been given a chance to have breakfast with the head monk, and I was hustling my way up a hill when a golf cart pulled up alongside me.

"Good morning," the man behind the wheel said to me. "Can I offer you a ride?"

Not wanting to keep Master Hsing Yun waiting, I accepted and climbed into the cart, telling the driver where I was headed. He was dressed in jeans and a simple long-sleeved shirt with an orange vest over it. He looked to be in his fifties like me, with streaks of grey in his hair. We rode in silence for a few minutes, absorbing the stillness of the landscape and the gentle breeze of cool morning air. As we rounded the hillside, I filled the silence with a bit of small talk.

"Do you do this for a living?" I asked.

"No," he replied. "I just volunteer here when I can find time outside my job."

I noticed that stitched across the left breast of his orange vest was the word "Volunteer" in Chinese characters.

"Well, what do you do for work?" I asked.

"I own an electronics manufacturing company and work as the CEO. But lately I've been spending less time working and more time volunteering. It's really special to see Master Hsing Yun sharing wisdom with people here. It brings a sense of serenity to help out with that in any way I can."

Those words, and the calm demeanor with which he spoke them, struck me. Electronics manufacturing can be a brutally competitive industry, one with razor-thin margins and unceasing pressure to innovate, upgrade, and optimize operations. Success often comes at the expense of health, with long hours at the factory bleeding into long nights drinking, smoking, and entertaining clients.

But the man driving the cart seemed healthy in body and totally at peace as he steered the golf cart up the winding path. He told me about how his weekends volunteering here at Fo Guang Shan had become a way to cleanse the burden and stress of his work week. He wasn't yet ready to retire, but the act of serving those who visited Fo Guang Shan let him tap into something both simpler and more profound than the machinations of his company.

When we reached Master Hsing Yun's quarters, I thanked the driver, and he replied with a nod of his head and a smile. During the breakfast that followed, the wisdom shared by Master Hsing Yun would have a profound impact on how I thought about my work and my life. But the conversation with the volunteer driving the golf cart also stayed with me.

At first, I thought his devotion to humbly serving those around him was something unique to the monastery, a function of the power of religious faith to unite and inspire us. But when I returned to Taipei for my medical treatment, I began to notice people wearing those orange volunteer vests all around the city: in the library, at busy traffic intersections, in county offices, and at national parks. They held up stop signs for children crossing the street, told park visitors about the indigenous flora of Taiwan, and guided people through the process of applying for health insurance. Many of the volunteers were elderly people or recently retired. Their pension plans took care of basic necessities, and so they devoted their time to helping others and maintaining solid bonds with their community.

As I underwent chemotherapy and began to contemplate the coming crises of the AI age, I often thought of the volunteers. While many individuals these days pontificate about using UBI as an all-purpose social sedative, I saw a certain wisdom in the humble activities of these volunteers and the broader communal culture they were creating. The city could, of course, go on functioning without this army of orange-vested, grey-haired volunteers . . . but it would feel a little less kind and a little less human. In that subtle transformation, I began to see a way forward.

THE SOCIAL INVESTMENT STIPEND: CARE, SERVICE, AND EDUCATION

Just as those volunteers devoted their time and energy toward making their communities a little bit more loving, I believe it is incumbent on us to use the economic abundance of the AI age to foster these same values and encourage this same kind of activity. To do this, I propose we explore the creation not of a UBI but of what I call a *social investment stipend*. The stipend would be a decent government salary given to those who invest their time and energy in those activities that promote a kind, compassionate, and creative society. These would include three broad categories: care work, community service, and education.

These would form the pillars of a new social contract, one that valued and rewarded *socially beneficial* activities in the same way we currently reward *economically productive* activities. The stipend would not substitute for a social safety net — the traditional welfare, healthcare, or unemployment benefits to meet basic needs — but would offer a respectable income to those who choose to invest energy in these socially productive activities. Today, social status is still largely tied to income and career advancement. Endowing these professions with respect will require paying them a respectable salary and offering the opportunity for advancement like a normal career. If executed well, the social investment stipend would nudge our culture in a more compassionate direction. It would put the economic bounty of AI to work in building a better society, rather than just numbing the pain of AI-induced job losses.

Each of the three recognized categories — care, service, and education — would encompass a wide range of activities, with different levels of compensation for full- and part-time participation. Care work could include parenting of young children, attending to an aging parent, assisting a friend or family member dealing with illness, or helping someone with mental or physical disabilities live life to the fullest. This category would create a veritable army of people — loved ones, friends, or even strangers — who could assist those in need, offering them what my entrepreneur friend's touchscreen device for the elderly never could: human warmth.

Service work would be similarly broadly defined, encompassing much of the current work of nonprofit groups as well as the kinds of volunteers I saw in Taiwan. Tasks could include performing environmental remediation, leading afterschool programs, guiding tours at national parks, or collecting oral histories from elders in our communities. Participants in these programs would register with an established group and commit to a certain number of hours of service work to meet the requirements of the stipend.

Finally, education could range from professional training for the jobs of the AI age to taking classes that could transform a hobby into a career. Some recipients of the stipend will use that financial freedom to pursue a degree in machine learning and use it to find a high-paying job. Others will use that same freedom to take acting classes or study digital marketing.

Bear in mind that requiring participation in one of these activities is not something designed to dictate the daily activities of each person receiving the stipend. That is, the beauty of human beings lies in our diversity, the way we each bring different backgrounds, skills, interests, and eccentricities. I don't seek to smother that diversity with a command-and-control system of redistribution that rewards only a narrow range of socially approved activities.

But by requiring some social contribution in order to receive the stipend, we would foster a far different ideology than the laissez-faire individualism of a UBI. Providing a stipend in exchange for participation in prosocial activities reinforces a clear message: It took efforts from people all across society to help us reach this point of economic abundance. We are now collectively using that abundance to recommit ourselves to one another, reinforcing the bonds of compassion and love that make us human.

Looking across all the activities, I believe there will be a wide enough range of choices to offer something suitable to all workers who have been displaced by AI. The more people-oriented may opt for care work, the more ambitious can enroll in job-training programs, and those inspired by a social cause may take up service or advocacy jobs.

In an age in which intelligent machines have supplanted us as the cogs and gears in the engine of our economy, I hope that we will value *all* of these pursuits — care, service, and personal cultivation — as part of our collective social project of building a more human society.

OPEN QUESTIONS AND SERIOUS COMPLICATIONS

Implementing a social investment stipend will of course raise new questions and frictions: How much should the stipend be? Should we reward people differently based on their performance in these activities? How do we know if someone is dutifully performing their "care" work? And what kinds of activities should count as "service" work? These are admittedly difficult questions, ones for which there are no clear-cut answers. Administering a social investment stipend in countries with hundreds of millions of people will involve lots of paperwork and legwork by governments and the organizations that create these new roles.

But these challenges are far from insurmountable. Governments in developed societies already attend to a dizzying array of bureaucratic tasks just to maintain public services, education systems, and social safety nets. Our governments already do the work of inspecting buildings, accrediting schools, offering unemployment benefits, monitoring sanitary conditions at hundreds of thousands of restaurants, and providing health insurance to tens of millions of people. Operating a social investment stipend would add to this workload, but I believe it would be more than manageable. Given the huge human upside to providing such a stipend, I believe the added organizational challenges will be well worth the rewards to our communities.

But what about affordability? Offering a living salary to people performing all of the above tasks would require massive amounts of revenue, totals that today appear unworkable in many heavily indebted countries. AI will certainly increase productivity across society, but can it really generate the huge sums necessary to finance such dramatic expansion in government expenditures?

This too remains an open question, one that will only be settled once the AI technologies themselves proliferate across our economies. If AI meets or exceeds predictions for productivity gains and wealth creation, I believe we could fund these types of programs through super taxes on super profits. Yes, it would somewhat cut into economic incentives to advance AI, but given the dizzying profits that will accrue to the winners in the AI age, I don't see this as a substantial impediment to innovation.

But it will take years to get to that place of astronomical profits, years during which working people will be hurting. To smooth the transition, I propose a slow ratcheting up of assistance. While leaping straight into the full social investment stipend described above likely won't work, I do think we will be able to implement incremental policies along the way. These piecemeal policies could both counteract job displacement as it happens and move us toward the new social contract.

We could start by greatly increasing government support for new parents so that they have the choice to remain at home or send their child to full-time daycare. For parents who choose to home-school their kids, the government could offer subsidies equivalent to a teacher's pay for those who attain certain certifications. In the public school systems, the number of teachers could also be greatly expanded — potentially by a factor as high as ten — with each teacher tasked with a smaller number of students that they can teach in concert with AI education programs. Government subsidies and stipends could also go to workers undergoing job retraining and people caring for aging parents. These simple programs would allow us to put in place the first building blocks of a stipend, beginning the work of shifting the culture and laying the groundwork for further expansion.

As AI continues to generate both economic value and worker displacement, we could slowly expand the purview of these subsidies to activities beyond care work or job training. And once the full impact of AI — very good for productivity, very bad for employment — becomes clear, we should be able to muster the resources and public will to implement programs akin to the social investment stipend.

When we do, I hope that this will not just alleviate the economic, social, and psychological suffering of the AI age. Rather, I hope that it will further empower us to live in a way that honors our humanity and empowers us to do what no machine can: share our love with those around us.

LOOKING FORWARD AND LOOKING AROUND

The ideas laid out in this chapter are an early attempt to grapple with the massive disruptions on the horizon of our AI future. We looked at technical fixes that seek to smooth the transition to an AI economy: retraining workers, reducing work hours, and redistributing income through a UBI. While all of these technical fixes have a role to play, I believe something more is needed. I envision the private sector creatively fostering human-machine symbiosis, a new wave of impact investing funding human-centric service jobs, and the government filling the gaps with a social investment stipend that rewards care, service, and education. Taken together, these would constitute a re-alignment of our economy and a rewriting of our social contract to reward socially productive activities.

These are not an exhaustive list or authoritative judgment on the ways in which we can adapt to widespread automation. But I do hope they provide at least a framework and a set of values to guide us in that process. Much of that framework comes from my under-standing of artificial intelligence and the global technology industry.

The values guiding these recommendations, however, are rooted in something far more intimate: the experience of my cancer diagno-sis and the personal transformation inspired by people like my wife, Master Hsing Yun, and so many others who selflessly shared their love and wisdom with me.

Had I never undergone that terrifying but ultimately enlighten-ing experience, I may never have woken up to the centrality of love in the human experience. Instead of seeking ways to foster a more loving and compassionate world, I would likely view the looming cri-ses through the same lens as those who are deep into AI today—as a simple resource-allocation problem to be dealt with in the most effi-cient way possible, likely through a UBI. It is only after going through my own personal trial by fire that I now see the hollowness of that approach.

My experience with cancer also taught me to appreciate the wis-dom that hides in the humble actions of people everywhere. After so many years as an "Ironman" of professional achievement, I needed to be knocked off my pedestal and face my own mortality before I appreciated what many so-called less successful people brought to the table.

I believe we will soon witness the same process on an interna-tional scale. The AI superpowers of the United States and China may be the countries with the expertise to build these technologies, but the paths to true human flourishing in the AI age will emerge from people in all walks of life and from all corners of the world.

As we look forward into the future, we must also take the time to look around.

9

★

OUR GLOBAL AI STORY

On June 12, 2005, Steve Jobs stepped up to a microphone in Stanford Stadium and delivered one of the most memorable commencement speeches ever given. In the talk, he retraced his zig-zagging career, from college dropout to cofounder of Apple, from his unceremonious ouster at that company to his founding of Pixar, and finally his triumphant return to Apple a decade later. Speaking to a crowd of ambitious Stanford students, many of whom were eagerly plotting their own ascent to the peaks of Silicon Valley, Jobs cautioned against trying to chart one's life and career in advance.

"You can't connect the dots looking forward," Jobs told the assembled students. "You can only connect them looking backwards. So you have to trust that the dots will somehow connect in your future."

Jobs's wisdom has resonated with me since I first heard it, but never more so than today. In writing this book, I've had the chance to connect the dots on four decades of work, growth, and evolution. That journey has spanned companies and cultures, from AI researcher and business executive to venture capitalist, author, and cancer survivor. It has touched on issues both global and deeply personal: the rise of artificial intelligence, the intertwined fates of the places that I've called home, and my own evolution from a workaholic to a more loving father, husband, and human being.

All of these experiences have come together to shape my view of our global AI future, to connect the dots looking backward and to use those constellations as guidance going forward. My background in technology and business expertise has crystallized how these technologies are developing in both China and the United States. My sudden confrontation with cancer woke me up to why we must use these technologies to foster a more loving society. Finally, my experience moving and transitioning between two different cultures has impressed on me the value of shared progress and the need for mutual understanding across national borders.

AN AI FUTURE WITHOUT AN AI RACE

In writing about global development of artificial intelligence, it's easy to revert to military metaphors and a zero-sum mentality. Many compare the "AI race" of today to the space race of the 1960s or, even worse, to the Cold War arms race that created ever more powerful weapons of mass destruction. Even the title of this book employs the word "superpowers," a phrase that many associate with geopolitical rivalry. I use this phrase, however, specifically to reflect the technological balance of AI capabilities, not to suggest an all-out struggle for military supremacy. But these distinctions are easily blurred by those more interested in political posturing than in human flourishing.

If we are not careful, this single-minded rhetoric around an "AI race" will undermine us in planning and shaping our shared AI future. A race has only one winner: China's gain is America's loss, and vice-versa. There is no notion of shared progress or mutual prosperity — just a desire to stay ahead of the other country, regardless of the costs. This mentality has led many commentators in the United States to use China's AI progress as a rhetorical whip with which to spur American leaders to action. They argue that America is at risk of losing its edge in the technology that will fuel the military competition of the twenty-first century.

But this is not a new Cold War. AI today has numerous potential military applications, but its true value lies not in destruction but in creation. If understood and harnessed properly, it can truly help all of us generate economic value and prosperity on a scale never before seen in human history.

In this sense, our current AI boom shares far more with the dawn of the Industrial Revolution or the invention of electricity than with the Cold War arms race. Yes, Chinese and American companies will compete with each other to better leverage this technology for productivity gains. But they are not seeking the conquest of the other nation. When Google promotes its TensorFlow technology abroad, or Alibaba implements its City Brain in Kuala Lumpur, these actions are more akin to the early export of steam engines and lightbulbs than as an opening volley in a new global arms race.

A clear-eyed look at the technology's long-term impact has revealed a sobering truth: in the coming decades, AI's greatest potential to disrupt and destroy lies not in international military contests but in what it will do to our labor markets and social systems. Appreciating the momentous social and economic turbulence that is on our horizon should humble us. It should also turn our competitive instincts into a search for cooperative solutions to the common challenges that we all face as human beings, people whose fates are inextricably intertwined across all economic classes and national borders.

GLOBAL WISDOM FOR THE AI AGE

As both the creative and disruptive force of AI is felt across the world, we need to look to each other for support and inspiration. The United States and China will lead the way in economically productive applications of AI, but other countries and cultures will certainly continue to make invaluable contributions to our broader social evolution. No single country will have all the answers to the tangled web of issues we face, but if we draw on diverse sources of wisdom, I believe there is no problem that we can't tackle together. This wisdom will include pragmatic reforms to our education systems, subtle nuances in cultural values, and deep shifts in how we conceive of development, privacy, and governance.

In revamping our education systems, we can learn much from

South Korea's embrace of gifted and talented education. These programs seek to identify and realize the potential of the country's top technical minds, an approach suited to creating the material prosperity that can then be broadly shared across society. Schools around the globe can also draw lessons from American experiments in social and emotional education, fostering skills that will prove invaluable to the human-centric workforce of the future.

For adaptations in how we approach work, we would be wise to look to the culture of craftsmanship in Switzerland and Japan, places where the pursuit of perfection has elevated routine work activities into the realm of human expression and artistry. Meanwhile, vibrant and meaningful cultures of volunteering in countries like Canada and the Netherlands should inspire us to diversify our traditional notions of "work." Chinese culture can also be a source of wisdom when it comes to caring for elders and in fostering intergenerational households. As public policy and personal values blend, we should really take the time to study new experiments in defining and measuring progress, such as Bhutan's decision to pursue "Gross National Happiness" as a key development indicator.

Finally, our governments will need to consistently look to one another in evaluating thorny new tradeoffs in data privacy, digital monopolies, online security, and algorithmic bias. In tackling these issues, we can learn much from comparing the different approaches taken by regulators in Europe, the United States, and China. While Europe has opted for a more heavy-handed approach (fining Google, for example, for antitrust and trying to wrest control over data away from the technology companies), China and the United States have given these companies greater leeway, letting technology and markets develop before intervening on the margins.

All these approaches present tradeoffs, with some favoring privacy over technological progress, and others doing the reverse. Leveraging technology to build the kind of societies we desire will mean following the real-world impact of these policies across geographies and remaining open-minded about different approaches to AI governance.

But accessing and embracing these diverse sources of insight first requires we maintain a sense of agency in relation to this quickly accelerating technology. With the daily barrage of headlines about AI, it's easy to feel as if human beings are losing control over our own destiny. Prophecies of both robot overlords and a "useless class" of unemployed workers tend to blend in our minds, conjuring up an overwhelming sense of human helplessness in the face of all-powerful technologies. Both of these doomsday scenarios contain a kernel of truth about AI's potential, but the feelings of helplessness they engender obscure the key point: when it comes to shaping the future of artificial intelligence, the single most important factor will be the actions of human beings.

We are not passive spectators in the story of AI — we are the authors of it. That means the values underpinning our visions of an AI future could well become self-fulfilling prophecies. If we tell ourselves that the value of human beings lies solely in their economic contribution, then we will act accordingly. Machines will displace humans in the workplace, and we may end up in a twisted world like the one Hao Jingfang imagined in *Folding Beijing*, a caste-based society that divides and separates the so-called useful people from the "useless" masses.

But this is in no way a foregone conclusion. The ideology underlying this dystopian vision — of human beings as nothing more than the sum of their economically productive parts — reveals just how far we've led ourselves astray. We were not put on Earth to merely grind away at repetitive tasks. We don't need to spend our lives busily accumulating wealth just so that we can die and pass it on to our children — the latest "iteration" of the human algorithm — who will refine and repeat that process.

If we believe that life has meaning beyond this material rat race, then AI just might be the tool that can help us uncover that deeper meaning.

When I launched my AI career in 1983, I did so by waxing philosophic in my application to the Ph.D. program at Carnegie Mellon. I described AI as "the quantification of the human thinking process, the explication of human behavior," and our "final step" to understanding ourselves. It was a succinct distillation of the romantic notions in the field at that time and one that inspired me as I pushed the bounds of AI capabilities and human knowledge.

Today, thirty-five years older and hopefully a bit wiser, I see things differently. The AI programs that we've created have proven capable of mimicking and surpassing human brains at many tasks. As a researcher and scientist, I'm proud of these accomplishments. But if the original goal was to truly understand myself and other human beings, then these decades of "progress" got me nowhere. In effect, I got my sense of anatomy mixed up. Instead of seeking to outperform the human brain, I should have sought to understand the human heart.

It's a lesson that it took me far too long to learn. I have spent much of my adult life obsessively working to optimize my impact, to turn my brain into a finely tuned algorithm for maximizing my own influence. I bounced between countries and worked across time zones for that purpose, never realizing that something far more meaningful and far more human lay in the hearts of the family members, friends, and loved ones who surrounded me. It took a cancer diagnosis and the unselfish love of my family for me to finally connect all these dots into a clearer picture of what separates us from the machines we build.

That process changed my life, and in a roundabout way has led me back to my original goal of using AI to reveal our nature as human beings. If AI ever allows us to truly understand ourselves, it will not be because these algorithms captured the mechanical essence of the human mind. It will be because they liberated us to forget about optimizations and to instead focus on what truly makes us human: loving and being loved.

Reaching that point will require hard work and conscious

choices by all of us. Luckily, as human beings, we possess the free will to choose our own goals that AI still lacks. We can choose to come together, working across class boundaries and national borders to write our own ending to the AI story.

Let us choose to let machines be machines, and let humans be humans. Let us choose to simply use our machines, and more importantly, to love one another.

ACKNOWLEDGMENTS

First and foremost, I want to thank my collaborator, Matt Sheehan, who did a tremendous amount of work on this book under a very tight deadline. If you feel this book is fun and easy to read, or maybe find it rich in information, Matt deserves much of the credit. I was lucky to find a collaborator like Matt, someone with a deep understanding of China, the United States, technology, and writing.

I was talked into doing this book by my friend and agent John Brockman and his team. His belief in the urgency of the subject and my ability to uniquely contribute to the conversation first persuaded me to consider taking on this project. In hindsight, I think he was absolutely right.

I'd like to thank Rick Wolff, who decided to bet on an unproven topic based on my own conviction. He is an outstanding editor and worked wonders in bringing this book to market. It was tremendous fun working with Rick — and pushing each other to be the best we could be.

I also want to thank Erik Brynjolfsson, James Manyika, Jonathan Woetzel, Paul Triolo, Shaolan Hsueh, Chen Xu, Ma Xiaohong, Lin Qiling, Wu Zhuohao, Michael Chui, Yuan Li, Cathy Yang, Anita Huang, Maggie Tsai, and Laurie Erlam for helping to read early drafts and giving me valuable feedback.

Final thanks go to my family, who have tolerated my inattentiveness during the past six months. I cannot wait to return to their embrace, an embrace that sustains me and has taught me so much. This should be my last book for a while. Then again, I've told them that seven times before — hopefully they'll still buy it.

NOTES

1. CHINA'S SPUTNIK MOMENT

page

2   *atoms in the known universe:* "Go and Mathematics," in Wikipedia, s.v., "Legal Positions," https://en.wikipedia.org/wiki/Go_and_mathematics#Legal_positions.

3   *280 million Chinese viewers:* Cade Metz, "What the AI Behind AlphaGo Can Teach Us About Being Human," *Wired,* May 19, 2016, https://www.wired.com/2016/05/google-alpha-go-ai/.

4   *issued an ambitious plan:* Paul Mozur, "Beijing Wants A.I. to Be Made in China by 2030," *New York Times,* July 20, 2017, https://www.nytimes.com/2017/07/20/business/china-artificial-intelligence.html.

     making up 48 percent: James Vincent, "China Overtakes US in AI Startup Funding with a Focus on Facial Recognition and Chips," *The Verge,* February 2, 2018, https://www.theverge.com/2018/2/22/17039696/china-us-ai-funding-startup-comparison.

     first software program: Kai-Fu Lee and Sanjoy Mahajan, "The Development of a World Class Othello Program," *Artificial Intelligence* 43, no. 1 (April 1990): 21–36.

8   *to create Sphinx:* Kai-Fu Lee, "On Large-Vocabulary Speaker-Independent Continuous Speech Recognition," *Speech Communication* 7, no. 4 (December 1988): 375–379.

9   *profile in the* New York Times: John Markoff, "Talking to Machines: Progress Is Speeded," *New York Times,* July 6, 1988, https://www.nytimes.com/1988/07/06/business/business-technology-talking-to-machines-progress-is-speeded.html?mcubz=1.

     demolished the competition: ImageNet Large Scale Visual Recognition Challenge 2012, Full Results, http://image-net.org/challenges/LSVRC/2012/results.html.

11   *for over $500 million:* Catherine Shu, "Google Acquires Artificial Intelligence Startup for Over $500 Million," *TechCrunch,* January 26, 2014, https://techcrunch.com/2014/01/26/google-deepmind/.

13   *harnessing of electricity:* Shana Lynch, "Andrew Ng: Why AI is the New Electricity," The Dish (blog), *Stanford News,* March 14, 2017, https://news.stanford.edu/thedish/2017/03/14/andrew-ng-why-ai-is-the-new-electricity/.

18   *add $15.7 trillion:* Dr. Anand S. Rao and Gerard Verweij, "Sizing the Prize," PwC, June 27, 2017, https://www.pwc.com/gx/en/issues/analytics/assets/pwc-ai-analysis-sizing-the-prize-report.pdf.

2. COPYCATS IN THE COLISEUM

22 *The Cloner:* Gady Epstein, "The Cloner," *Forbes,* April 28, 2011, https://www.forbes.com/global/2011/0509/companies-wang-xing-china-groupon-friendster-cloner.html#1272f84055a6.

"*A Mark Zuckerberg Production*": 孙进, 李静颖 孙进, and 刘佳, "社交媒体冲向互联网巅峰," 第一财经日报, April 21, 2011, http://www.yicai.com/news/739256.html.

28 "*let some people get rich first*": "To Each According to His Abilities," *Economist,* May 31, 2001, https://www.economist.com/node/639652.

31 "*come to Shijingshan!*": Gabrielle H. Sanchez, "China's Counterfeit Disneyland Is Actually Super Creepy," BuzzFeed, December 11, 2014, https://www.buzzfeed.com/gabrielsanchez/chinas-eerie-counterfeit-disneyland.

33 *0.2 percent of the Chinese population:* Xueping Du, "Internet Adoption and Usage in China," 27th Annual Telecommunications Policy and Research Conference, Alexandria, VA, September 25–27, 1999, https://pdfs.semanticscholar.org/4881/o88c67ad919da32487c567341f8a0af7e47e.pdf.

36 "*free is not a business model*": "Ebay Lectures Taobao That Free Is Not a Business Model," *South China Morning Post,* October 21, 2005, http://www.scmp.com/node/521384.

42 *his autobiography,* Disruptor: 周鸿祎, "颠覆者" (北京: 北京联合出版公司, 2017).

44 *Sinovation event in Menlo Park:* Dr. Andrew Ng, Dr. Sebastian Thrun, and Dr. Kai-Fu Lee, "The Future of AI," moderated by John Markoff, Sinovation Ventures, Menlo Park, CA, June 10, 2017, http://us.sinovationventures.com/blog/the-future-of-ai.

book The Lean Startup: Eric Ries, *The Lean Startup: How Today's Entrepreneurs Use Continuous Innovation to Create Radically Successful Businesses* (New York: Crown Business, 2011).

3. CHINA'S ALTERNATE INTERNET UNIVERSE

58 *the Next Web:* Francis Tan, "Tencent Launches Kik-Like Messaging App," The Next Web, January 21, 2011, https://thenextweb.com/asia/2011/01/21/tencent-launches-kik-like-messaging-app-in-china/.

59 "*remote control for life*": Connie Chan, "A Whirlwind Tour Through China Tech Trends," Andreesen Horowitz (blog), February 6, 2017, https://a16z.com/2017/02/06/china-trends-2016-2017/.

60 "*Pearl Harbor attack*": Josh Horwitz, "Chinese WeChat Users Sent out 20 Million Cash-Filled Red Envelopes to Friends and Family Within Two Days," TechinAsia, February 4, 2014, https://www.techinasia.com/wechats-money-gifting-scheme-lures-5-million-chinese-users-alibabas-jack-ma-calls-pearl-harbor-attack-company.

62 "*mass entrepreneurship and mass innovation*": "Premier Li's Speech at Summer Davos Opening Ceremony," *Xinhua,* September 10, 2014, http://english.gov.cn/premier/speeches/2014/09/22/content_281474988575784.htm.

64   *nearly quadrupling:* Zero2IPO Research，"清科观察：《2016政府引导基金报
告》发布，管理办法支持四大领域、明确负面清单，" 清科研究中心，March
30, 2016, http://free.pedata.cn/1440998436840710.html.

65   *quadrupled to $12 billion:* "Venture Pulse Q4 2017," *KPMG Enterprise,* January
16, 2018, https://assets.kpmg.com/content/dam/kpmg/xx/pdf/2018/01/ven-
ture-pulse-report-q4-17.pdf.

69   *ten times the total:* Thomas Laffont and Daniel Senft, "East Meets West 2017
Keynote," East Meets West 2017 Conference, Pebble Beach, CA, June 26–29,
2017.

72   *"do what we do best":* Joshua Brustein, "GrubHub Buys Yelp's Eat24 for $288
Million," *Bloomberg,* August 3, 2017, https://www.bloomberg.com/news/arti-
cles/2017-08-03/grubhub-buys-yelp-s-eat24-for-288-million.

73   *study by McKinsey and Company:* Kevin Wei Wang, Alan Lau, and Fang Gong,
"How Savvy, Social Shoppers Are Transforming Chinese E-Commerce," Mc-
Kinsey and Company, April 2017, https://www.mckinsey.com/industries/
retail/our-insights/how-savvy-social-shoppers-are-transforming-chinese
-e-commerce.

75   *753 million smartphone users:* 第41次 "中国互联网络发展状况统计报告,"
中国互联网络信息中心，January 18, 2018, http://www.cac.gov.cn/2018=01
/31/c_1122346138.htm.
"no cash left in Hangzhou?": "你的城市还用现金吗? 杭州的劫匪已经抢不到
钱了," 吴晓波频道, April 3, 2017, http://www.sohu.com/a/131836799_565426.
iResearch estimated in 2017: "China's Third-Party Mobile Payments Report,"
iResearch, June 28, 2017, http://www.iresearchchina.com/content/de-
tails8_34116.html.
surpassed $17 trillion: Analysis 易观，"中国第三方支付移动支付市场季
度监测报告2017年第4季度," http://www.analysis.cn/analysis/trade/de-
tail/1001257/.

78   *for $2.7 billion:* Cate Cadell, "China's Meituan Dianping Acquires Bike-Shar-
ing Firm Mobike for $2.7 Billion," *Reuters,* April 3, 2018, https://www.reuters.
com/article/us-mobike-m-a-meituan/chinas-meituan-dianping-acquires-
bike-sharing-firm-mobike-for-2-7-billion-idUSKCN1HB0DU.

79   *three hundred to one:* Laffont and Senft, "East Meets West 2017 Keynote."

4. A TALE OF TWO COUNTRIES

89   *"put AAAI on Christmas day":* Sarah Zhang, "China's Artificial Intelligence
Boom," *Atlantic,* February 16, 2017, https://www.theatlantic.com/technology/
archive/2017/02/china-artificial-intelligence/516615/.
23.2 percent to 42.8: Dr. Kai-Fu Lee and Paul Triolo, "China Embraces AI: A
Close Look and a Long View," presentation at Eurasia Group, December 6,
2017, https://www.eurasiagroup.net/live-post/ai-in-china-cutting-through-
the-hype.
one hundred most-cited research institutions: Shigenori Arai, "China's AI Ambitions Revealed by List of Most Cited Research Papers," Nikkei Asian Review,
November 2, 2017, https://asia.nikkei.com/Tech-Science/Tech/China-s-AI-
ambitions-revealed-by-list-of-most-cited-research-papers.

90   *"these Chinese people are good"*: Same Shead, "Eric Schmidt on AI: 'Trust Me, These Chinese People Are Good,'" *Business Insider,* November 1, 2017, http://www.businessinsider.com/eric-schmidt-on-artificial-intelligence-china-2017-11.

93   *Google's own R&D budget:* Gregory Allen and Elsa B. Kania, "China Is Using America's Own Plan to Dominate the Future of Artificial Intelligence," *Foreign Policy,* September 8, 2017, http://foreignpolicy.com/2017/09/08/china-is-using-americas-own-plan-to-dominate-the-future-of-artificial-intelligence/.

      "historic achievement": Allison Linn, "Historic Achievement: Microsoft Researchers Reach Human Parity in Conversational Speech Recognition," The AI Blog, Microsoft, October 18, 2016, https://blogs.microsoft.com/ai/historic-achievement-microsoft-researchers-reach-human-parity-conversational-speech-recognition/.

      Ng left Baidu: Andrew Ng, "Opening a New Chapter of My Work in AI," Medium, March 21, 2017, https://medium.com/@andrewng/opening-a-new-chapter-of-my-work-in-ai-c6a4d1595d7b.

98   *proposed* cutting *funding:* Paul Mozur and John Markoff, "Is China Outsmarting America in A.I.?" *New York Times,* May 27, 2017, https://www.nytimes.com/2017/05/27/technology/china-us-ai-artificial-intelligence.html?_r=0.

100   *"venture socialism":* "Capitalizing on 'Venture Socialism,'" *Washington Post,* September 18, 2011, https://www.washingtonpost.com/opinions/capitalizing-on-venture-socialism/2011/09/16/gIQAQ7sYdK_story.html?utm_term=.5f0e532fcb86.

101   *260,000 annual road fatalities:* "Scale of Traffic Deaths and Injuries Constitutes 'a Public Health Crisis' — Safe Roads Contribute to Sustainable Development," World Health Organization, Western Pacific Region, press release, May 24, 2016, http://www.wpro.who.int/china/mediacentre/releases/2016/20160524/en/.

5. THE FOUR WAVES OF AI

108   *coined the famous phrase*: Frederick Jelinek, "Some of My Best Friends Are Linguists," presentation at the International Conference on Language Resources and Evaluation, May 28, 2004, http://www.lrec-conf.org/lrec2004/doc/jelinek.pdf.

      seventy-four minutes per day: "Toutiao, a Chinese News App That's Making Headlines," *Economist,* November 18, 2017, https://www.economist.com/news/business/21731416-remarkable-success-smartphone-app-claims-figure-users-out-within-24.

113   *"new standard of beauty":* Conversation with author, October 2017.

116   *rank all prosecutors:* 朱晓颖，"江苏"案管机器人"很忙: 辅助办案 还考核检察官," 中国新闻网, March 2, 2018, http://www.chinanews.com/sh/2018/03-02/8457963.shtml.

127   *85 million by the end of 2017:* Sarah Dai, "China's Baidu, Xiaomi in AI Pact to Create Smart Connected Devices," *South China Morning Post,* November 28, 2017, http://www.scmp.com/tech/china-tech/article/2121928/chinas-baidu-xiaomi-ai-pact-create-smart-connected-devices.

toward an IPO predicted: Shona Gosh, "Xiaomi Is Picking up Underwriters for an IPO Worth up to $100 Billion," *Business Insider*, January 15, 2018, http://www.businessinsider.com/xiaomi-goldman-sachs-ipo-100-billion-2018-1.

130 *"the best company"*: April Glaser, "DJI Is Running away with the Drone Market," Recode, April 14, 2017, https://www.recode.net/2017/4/14/14690576/drone-market-share-growth-charts-dji-forecast.

132 *1.5 million miles:* Fred Lambert, "Google's Self-Driving Car vs Tesla Autopilot: 1.5M Miles in 6 Years vs 47M Miles in 6 Months," *Electrek*, April 11, 2016, https://electrek.co/2016/04/11/google-self-driving-car-tesla-autopilot/.

133 *$583 billion:* "Xiong'an New Area: China's Latest Special Economic Zone?" CKGSB Knowledge, November 8, 2017, http://knowledge.ckgsb.edu.cn/2017/11/08/all-articles/xiongan-china-special-economic-zone/.

6. UTOPIA, DYSTOPIA, AND THE REAL AI CRISIS

141 *Kurzweil predicts:* Dom Galeon and Christianna Reedy, "Kurzweil Claims That the Singularity Will Happen by 2045," *Futurism,* October 5, 2017, https://futurism.com/kurzweil-claims-that-the-singularity-will-happen-by-2045/.

"the biggest risk we face": James Titcomb, "AI Is the Biggest Risk We Face as a Civilisation, Elon Musk Says," London *Telegraph*, July 17, 2017, https://www.telegraph.co.uk/technology/2017/07/17/ai-biggest-risk-face-civilisation-elon-musk-says/.

"summoning the demon": Greg Kumparak, "Elon Musk Compares Building Artificial Intelligence to 'Summoning the Demon,'" TechCrunch, October 26, 2014, https://techcrunch.com/2014/10/26/elon-musk-compares-building-artificial-intelligence-to-summoning-the-demon/.

142 *median prediction of 2040:* Nick Bostrom, *Superintelligence: Paths, Dangers, Strategies* (Oxford: Oxford University Press, 2014), 19.

143 *Hinton and his colleague's landmark paper:* Geoffrey Hinton, Simon Osindero, and Yee-Whye The, "A Fast Learning Algorithm for Deep Belief Nets," *Neural Computation* 18 (2006): 1527–1554.

144 *Folding Beijing:* Hao Jingfang, *Folding Beijing*, trans. Ken Liu, *Uncanny Magazine,* https://uncannymagazine.com/article/folding-beijing-2/.

147 *suffer stagnant wages:* Robert Allen, "Engel's Pause: A Pessimist's Guide to the British Industrial Revolution," University of Oxford Department of Economics Working Papers, April 2007, https://www.economics.ox.ac.uk/department-of-economics-discussion-paper-series/engel-s-pause-a-pessimist-s-guide-to-the-british-industrial-revolution.

148 *technologies that "really matter"*: Erik Brynjolfsson and Andrew McAfee, *The Second Machine Age: Work, Progress, and Prosperity in a Time of Brilliant Technologies* (New York: Norton, 2014), 75–77.

150 *"the great decoupling"*: Erik Brynjolfsson and Andrew McAfee, "Jobs, Productivity and the Great Decoupling," *New York Times,* December 11, 2012, http://www.nytimes.com/2012/12/12/opinion/global/jobs-productivity-and-the-great-decoupling.html.

doubled its share: Eduardo Porter and Karl Russell, "It's an Unequal World.

It Doesn't Have to Be," *New York Times,* December 14, 2017, https://www.ny-times.com/interactive/2017/12/14/business/world-inequality.html.

twice as much wealth: Matt Egan, "Record Inequality: The Top 1% Controls 38.6% of America's Wealth," CNN, September 17, 2017, http://money.cnn.com/2017/09/27/news/economy/inequality-record-top-1-percent-wealth/index.html.

fallen for the poorest Americans: Lawrence Mishel, Elise Gould, and Josh Bivens, "Wage Stagnation in Nine Charts," Economic Policy Institute, January 6, 2015, http://www.epi.org/publication/charting-wage-stagnation/.

151 *"The answer is surely not":* Claire Cain Miller, "As Robots Grow Smarter, American Workers Struggle to Keep Up," The Upshot (blog), *New York Times,* December 15, 2014, https://www.nytimes.com/2014/12/16/upshot/as-robots-grow-smarter-american-workers-struggle-to-keep-up.html.

"the biggest challenge": Ibid.

153 *$148 billion invested:* Dana Olsen, "A Record-Setting Year: 2017 VC Activity in 3 Charts," Pitchbook, December 15, 2017, https://pitchbook.com/news/articles/a-record-setting-year-2017-vc-activity-in-3-charts.

leapt to $15.2 billion: "Top AI Trends to Watch in 2018," *CB Insights,* February 2018, https://www.cbinsights.com/research/report/artificial-intelligence-trends-2018/.

158 *a dire prediction:* Carl Benedikt Frey and Michael A. Osborne, "The Future of Employment: How Susceptible Are Jobs to Automation," Oxford Martin Programme on Technology and Employment, September 17, 2013, https://www.oxfordmartin.ox.ac.uk/downloads/academic/future-of-employment.pdf.

just 9 percent of jobs: Melanie Arntz, Terry Gregory, and Ulrich Zierahn, "The Risk of Automation for Jobs in OECD Countries: A Comparative Analysis," *OECD Social, Employment, and Migration Working Papers,* no. 189, May 14, 2016, http://dx.doi.org/10.1787/5jlz9h56dvq7-en.

159 *38 percent of jobs:* Richard Berriman and John Hawksworth, "Will Robots Steal Our Jobs? The Potential Impact of Automation on the UK and Other Major Economies," PwC, March 2017, https://www.pwc.co.uk/economic-services/ukeo/pwcukeo-section-4-automation-march-2017-v2.pdf.

160 already automatable: James Manyika et al., "What the Future of Work Will Mean for Jobs, Skills, and Wages," McKinsey Global Institute, November 2017, https://www.mckinsey.com/global-themes/future-of-organizations-and-work/what-the-future-of-work-will-mean-for-jobs-skills-and-wages.

164 *20 to 25 percent fewer employees:* Karen Harris, Austin Kimson, and Andrew Schwedel, "Labor 2030: The Collision of Demographics, Automation and Inequality," Bain and Company, February 7, 2018, http://www.bain.com/publications/articles/labor-2030-the-collision-of-demographics-automation-and-inequality.aspx.

make China "ground zero": Martin Ford, "China's Troubling Robot Revolution," *New York Times,* June 10, 2015, https://www.nytimes.com/2015/06/11/opinion/chinas-troubling-robot-revolution.html.

165 *"American robots work as hard":* Vivek Wadhwa, "Sorry China, the Future

of Next-Generation Manufacturing Is in the US," *Quartz,* August 30, 2016, https://qz.com/769897/sorry-china-the-future-of-next-generation-manu facturing-is-in-the-us/.

169   *capture a full 70 percent:* Rao and Verweij, "Sizing the Prize."

172   *"useless class":* Yuval N. Harari, "The Rise of the Useless Class," TED Ideas, February 24, 2017, https://ideas.ted.com/the-rise-of-the-useless-class/.

173   *"I lost my sense of worth":* Binyamin Appelbaum, "The Vanishing Male Worker: How America Fell Behind," *New York Times,* December 11, 2014, https://www.nytimes.com/2014/12/12/upshot/unemployment-the-vanishing-male-worker-how-america-fell-behind.html.

      Rates of depression triple: Rebecca J. Rosen, "The Mental-Health Consequences of Unemployment," *Atlantic,* June 9, 2014, https://www.theatlantic.com/business/archive/2014/06/the-mental-health-consequences-of-unem ployment/372449/.

      "deaths of despair": Anne Case and Angus Deaton, "Mortality and Morbidity in the 21st Century," Brookings Papers on Economic Activity, Spring 2017, https://www.brookings.edu/wp-content/uploads/2017/08/casetextsp17b pea.pdf.

7. THE WISDOM OF CANCER

180   Be Your Personal Best: 李开复，做最好的自己 (北京: 人民出版社, 2005), https://www.amazon.cn/dp/B00116LOoW.

      Making a World of Difference: Dr. Kai-Fu Lee, Haitao Fan, and Crystal Tai (translator), *Making a World of Difference,* Amazon Digital Services, April 13, 2018.

187   *"It all comes down to love":* Bronnie Ware, "Top 5 Regrets of the Dying," *Huffington Post,* January, 21, 2012, https://www.huffingtonpost.com/bronnie-ware/top-5-regrets-of-the-dyin_b_1220965.html.

188   *five stages of grief:* Elisabeth Kübler-Ross, *On Death and Dying* (New York: Macmillan, 1969).

191   *analyzed fifteen different variables:* Massimo Federico et al., "Follicular Lymphoma International Prognostic Index 2: A New Prognostic Index for Follicular Lymphoma Developed by the International Follicular Lymphoma Prognostic Factor Project," *Journal of Clinical Oncology* 27, no. 27 (September 2009): 4555–4562.

8. A BLUEPRINT FOR HUMAN COEXISTENCE WITH AI

205   *move to a four-day work week:* Seth Fiegerman, "Google Founders Talk About Ending the 40-Hour Work Week," *Mashable,* July 7, 2014, https://mashable.com/2014/07/07/google-founders-interview-khosla/#tXe9XU.mr5qU.

      creative approaches to work-sharing: Steven Greenhouse, "Work-Sharing May Help Companies Avoid Layoffs," *New York Times,* June 15, 2009, http://www.nytimes.com/2009/06/16/business/economy/16workshare.html.

207   *Y Combinator president Sam Altman:* Kathleen Pender, "Oakland Group Plans to Launch Nation's Biggest Basic-Income Research Project," *San Francisco Chronicle,* September 21, 2017, https://www.sfchronicle.com/business/net

worth/article/Oakland-group-plans-to-launch-nation-s-biggest-12219073. php.

Facebook cofounder Chris Hughes: The Economic Security Project, https://economicsecurityproject.org/.

208 *gives a thousand families a stipend:* Pender, "Oakland Group."

"VC for the People": Steve Randy Waldman, "VC for the People," Interfluidity (blog), April 16, 2014, http://www.interfluidity.com/v2/5066.html.

"everyone has a cushion": Chris Weller, "Mark Zuckerberg Calls for Exploring Basic Income in Harvard Commencement Speech," *Business Insider,* May 25, 2017, http://www.businessinsider.com/mark-zuckerberg-basic-income-harvard-speech-2017-5.

214 *two fastest growing professions:* Ben Casselman, "A Peek at Future Jobs Reveals Growing Economic Divides," *New York Times,* October 24, 2017, https://www.nytimes.com/2017/10/24/business/economy/future-jobs.html.

just over $20,000: U.S. Department of Labor, Bureau of Labor Statistics, Occupational Employment Statistics, "Home Health Aides and Personal Care Aides," https://www.bls.gov/ooh/healthcare/home-health-aides-and-personal-care-aides.htm, and "Personal Care Aides," https://www.bls.gov/oes/current/oes399021.htm.

215 *"A Sense of Purpose":* Larry Fink, "Larry Fink's Annual Letter to CEOs: A Sense of Purpose," BlackRock, January 18, 2018, https://www.blackrock.com/corporate/en-us/investor-relations/larry-fink-ceo-letter.

9. OUR GLOBAL AI STORY

226 *"You can't connect the dots":* Steve Jobs, "2005 Stanford Commencement Address," Stanford University, published March 7, 2018, https://www.youtube.com/watch?v=UF8uR6Z6KLc&t=785s.

227 *space race of the 1960s:* John R. Allen and Amir Husain, "The Next Space Race Is Artificial Intelligence: And the United States Is Losing," *Foreign Policy,* November 3, 2017, http://foreignpolicy.com/2017/11/03/the-next-space-race-is-artificial-intelligence-and-america-is-losing-to-china/.

Cold War arms race: Zachary Cohen, "US Risks Losing Artificial Intelligence Arms Race to China and Russia," CNN, November 29, 2017, https://www.cnn.com/2017/11/29/politics/us-military-artificial-intelligence-russia-china/index.html.

INDEX

ABOUT THE AUTHOR

Dr. Kai-Fu Lee is the chairman and CEO of Sinovation Ventures and the president of Sinovation Ventures' Artificial Intelligence Institute. Sinovation, which manages $1.7 billion in dual-currency investment funds, is a leading venture capital firm focused on developing the next generation of Chinese high-tech companies.

Before founding Sinovation in 2009, Lee was the president of Google China. He previously held executive positions at Microsoft, SGI, and Apple. Lee received his bachelor's degree in computer science at Columbia University and his Ph.D. from Carnegie Mellon University. He holds honorary doctoral degrees from Carnegie Mellon and the City University of Hong Kong and is a fellow of the Institute of Electrical and Electronics Engineers (IEEE). Lee is the author of seven best-selling books in China.

In the field of artificial intelligence, Lee founded Microsoft Research China, which was named the "hottest computer lab" by *MIT Technology Review*. Later renamed Microsoft Research Asia, this institute trained the great majority of AI leaders in China, including CTOs or AI heads at Baidu, Tencent, Alibaba, Lenovo, Huawei, and Haier. While at Apple, Lee led AI projects in speech and natural language, which have been featured on *Good Morning America* and the front page of the *Wall Street Journal*. He is the author of ten U.S. patents and more than one hundred journal and conference papers. Altogether, Lee has been in artificial intelligence research, development, and investment for more than thirty years.

For more information on Kai-Fu Lee, visit www.aisuperpowers .com or follow him on Twitter: @kaifulee.